Beyond DNA

Inheriting Spiritual Strength from the Women in Your Family Tree

Selena Post

Selena Post

CROSSBOOKS
PUBLISHING

CrossBooks™
A Division of LifeWay
1663 Liberty Drive
Bloomington, IN 47403
www.crossbooks.com
Phone: 1-866-879-0502

First published by CrossBooks 07/15/2011

ISBN: 978-1-4627-0503-0 (sc)
ISBN: 978-1-4627-0504-7 (hc)

Library of Congress Control Number: 2011932341

Printed in the United States of America

This book is printed on acid-free paper.

for

Gloria Inez Hammer Cook

World's Best Sunday School Teacher

Charm can deceive and beauty fades.

A woman who loves God will be remembered and praised.

The essence of Proverbs 31:30

Contents

Introduction

God's mercy is certain for those in every generation who reverence him.

Based on Mary's words in Luke 1:50

March 2011: Reports of terrorist attacks, war, and natural disasters bombard us from newspaper pages and television screens. Stories of violent gang activity, drug-related crime, child abuse, and animal cruelty are frequent topics of local news broadcasts. Confessions of morally bankrupt religious leaders, political officials, and sports figures shock and disappoint us. Unsettling financial information streams across our monitors. "These are the worst of times," we might say. Are these truly the worst of times?

People of every era face challenges. We are alive today because each generation prevailed to create the next. Triumph over adversity could illustrate physical strength, although our ancestors needed more than brute force. Did their spiritual strength carry them through? The world's issues can influence our physical lives; however, inner peace is ours to control. We can allow the turmoil of daily life to distract us, or we can focus on positive action. With God's help, all people can become spiritually strong.

Women play a vital role because they operate in positions of great power. Women literally bring the next generation into the world. Strategically positioned, women instill faith, morals, and wisdom in their children. Natural communicators, women share ideas influentially with family members and friends. Imagine vastly increased numbers of women studying God's word, praying, and becoming spiritually strong. Envision an army of spiritually strong women modeling and teaching those practices among people in their spheres of influence. Imagine the resulting decline in crime, family discord, substance abuse, poverty, and general unhappiness. We can do more than imagine.

How do we become spiritually strong women who can and will change the world for good? We know we need to eat properly, sleep well, and exercise to be physically strong. Specific spiritual preparations are even more important. We must be armed and dangerous (in a good way). We arm ourselves by putting on the armor of God (by knowing His Word and mirroring His character). When properly armed to protect our families and ourselves, we become dangerous to God's enemies and to concepts that challenge God's plan. We can certainly begin by building and using our spiritual strength on our own, although we'll gain greater insight and encouragement by identifying and observing spiritually strong female role models.

Looking to other women for help is a natural process for women. We are quick to ask other women advice when we're choosing doctors and hairdressers. In grocery stores, we ask strangers for product advice. We poll our friends when choosing a major appliance. We read biographies of women we admire. Similarly, we can learn to apply our spiritual strength in life's difficult moments by observing spiritually strong women. Where might we find spiritual role models? They're sitting among the branches of our own family trees.

First, let's clarify some basics. What do I mean by *spiritual*? There are many forms of spirituality in our twenty-first century world. The perspective I present is the spiritual definition my ancestors understood: Christianity. Beyond that foundational choice, my ancestors worshiped in several Christian denominations. Like me, some women in my ancestry worshiped in more than one denomination in their lifetime. Perhaps I inherited their desire to seek God's best. What is imperative about our denominational choice? We must worship the one true God, and our practice of worship must conform to the teachings of the Bible. Our focus should be on our relationship with God rather than on rituals of religion.

May women who are not Christians participate? Our God the Father is the same God who, for example, our Jewish friends worship. Jewish women can mirror our process of investigating the lives of women in their families to discover spiritual evidence of their own faith. Although people of any faith might benefit from the process of researching the spiritual lives of their ancestors, this book espouses the Christian perspective I share with my ancestors.

How do we *inherit* spiritual strength? In practical terms, we cannot inherit our spiritual conversion from another person. Each person must ask to receive salvation through Jesus Christ. People of faith can, however, lead us toward God's saving grace through their example, instruction, and nurture. As spiritual babes, we accelerate our crawling-to-standing-to-walking process by emulating seasoned believers who model the benefits of worship, Bible study, and prayer.

What are the characteristics of a spiritually strong woman? She stands on God's Word to see beyond today's problems. In the power of the Holy Spirit, she maintains a positive posture of hope. She prays without ceasing (she is in continuous communication with God because she has accepted Jesus as her Savior and intercessor). A spiritually strong woman invests her best energy in people (not things) where dividends are eternal. She shares her faith. She strives to align her life with Biblical teachings. Because she is human, she is not perfect; she fails, yet she never gives up.

Adopted women are included in our adventure. One of the genealogies for Jesus (see Luke 3) mentions Jesus was *known* as Joseph's son rather than *being* his son. Matthew's version refers to Joseph as the husband of Mary, mother of Jesus, yet still treats Joseph's ancestry as pertinent. An adopted woman can focus on her adoptive ancestral women's teaching and nurture as an inheritance. Adoptive mothers and grandmothers who model and teach spiritual truths become spiritual ancestral mothers.

Why study the spiritual lives of women to whom we're related by DNA or nurture? Beyond widely applauded women of our own era and women whose stories we read in scripture, there are scores of women whose quiet lives of faith positively influenced society throughout history. Among historical female strangers, we find many excellent role models. It could be difficult, however, to identify closely with strangers whose lives and heritage are quite different from our own. How can we find and investigate women whose lives are more relevant to us? To explain the process, I'll begin with my own spiritual journey.

My early spiritual training began with my mother's instruction and continued under the ministry of Presbyterian churches in New York City and Denver. Arriving in North Carolina in 1955, I began to learn of God's plan under the loving ministry of Bonlee Baptist Church. By 1957, I hungered for salvation but was too shy to walk down the church aisle to publicly profess my faith. When I mentioned my dilemma to my mother's Quaker sister, Carol, God provided a way. Aunt Carol explained that Jesus would meet me any place, any time. That night, I received salvation by my bedside at the age of ten. Fifty-three years later, I'm still grateful. It was, and still is, the best decision of my life.

When my mother refused to attend church with me, women in my church family graciously stepped in to nurture my faith. Miss Ina, Inez, Lois, and Polly were foremost among the many women who became my spiritual mothers. My maternal grandmother, Lena, read her Bible and prayed in my presence. Lena, Aunt Carol, and my church family helped change my life for good. On my eighteenth birthday, Lena was present when I walked down our church aisle, professed my faith, requested baptism, and became a church member. Since that moment on my knees at age ten, God has been fully available as my tower of strength. How else does one survive life?

Knowing faith to be vital to my own well-being, I wondered about the women with whom I share DNA. While seeking evidence of their faith, I also wanted to know everything possible about their lives so I'd have a measure of the difficulties they endured. I wanted to find churches where they worshiped, learn the names of pastors they heard, and understand the ministries of their denominations. Could I learn whether they read their Bibles and were prayer warriors? Could I determine whether their faith was a positive influence in their lives and in the lives of their peers and descendants?

You'll learn the answers to those questions as you read the stories of fifteen women in my own ancestry: my mother, two grandmothers, four great-grandmothers, and eight great, great-grandmothers. Noteworthy women of earlier generations make brief appearances in the stories of their descending daughters. You might find the stories of my women compelling, yet stories of your own women will be even more pertinent to you. If your experience parallels mine, you will learn of difficulties they endured and sacrifices they made. Perhaps you'll shed tears when you discover their sorrows. You'll love them though you might not meet them this side of heaven. Their endurance and sacrifice ultimately gave you life. Beyond life itself, you might find their physical characteristics, skills, or interests live on in you.

Where and how should you begin your research? *Genealogy* is a term used to describe the actual chart illustrating one's pedigree (or line of ancestors) as well as the process for developing such charts. Identifying our ancestors and organizing information about their lives is the foundational step on which we can build specialized research projects such as our quest for knowledge about the spiritual strengths of our ancestral women.

What does the Bible say about genealogy? By the fourth chapter of Genesis in the Old Testament and in the opening verses of the first chapter of Matthew in the New Testament, we find lists of related people. There are lengthy and abbreviated family trees in scripture and some Biblical genealogies include the names of women. My favorite is a brief genealogy in the closing verses of the book of Ruth. That passage mentions the male ancestors of Ruth's second husband then ends with Ruth's son, grandson, and great-grandson, King David. One verse, 1 Timothy 1:4, might initially sound negative. The *endless genealogies* mentioned in the Timothy verse refer to the practice of falsely embellishing valid genealogies of Old Testament scripture. Christian women never falsely embellish a family history.

People new to genealogy benefit by taking a course, attending a genealogy conference, reading books on the subject, physically or virtually visiting libraries, and searching instructive genealogical websites. Although not intended as an instructional text on basic genealogical research, this book includes descriptions of processes that worked well for me when facts were illusive. You'll also learn how I connected with generous people who provided information, photographs, and encouragement.

Family research is simultaneously enthralling, addictive, frustrating, free, expensive, and fulfilling. Are you inquisitive by nature? Do you enjoy projects that could last years or the rest of your life? If not, find a relative willing to perform the research and document the findings. You could also hire a professional genealogist. I suggest, however, that you not be too quick to miss an opportunity to become a family detective. You might find the process even more rewarding than the results. Researching the lives of my ancestors is the most fulfilling adventure of my life. Genealogy with a spiritual focus is even more thrilling than my previously unsurpassed, much-loved, thirty-year career in aerospace supporting manned space flight.

Seasoned genealogists with fully developed family trees will find investigating spiritual history a welcome addition to traditional research. By adding what we learn about each woman's faith to her personal story, we can better understand how she endured life's challenges. You could unearth inspiring family stories of your ancestors exhibiting faith during history's most difficult moments. Even better, you might find documentary evidence penned by the women themselves.

When a woman's spiritual story eludes you, seek the most universally available facts. Her grave might be in a cemetery adjacent to the church where she worshiped during her later life. Search for christening or baptismal records of the woman, her parents, and her children. Look on her marriage license or newspaper announcement for the name and church affiliation of the minister who performed her marriage. Local church histories provide insights about religious education and worship opportunities available during your ancestor's lifetime.

Many women you'll meet in this book, and their associated families, worshiped in Methodist churches. Formal names of Methodist churches have undergone modification. Most of the churches I mention remain open and vital today as part of the United Methodist Church. For simplicity, I use the term *Methodist* coupled with identifying church names such as Hope, Bethlehem, Centenary, and Larchmont.

Genealogy offers a life-expanding experience for people who usually achieve goals alone. As an only child, I learned early to enjoy solitude and began my research solo. My advice is to involve as many people as possible in your project. When formal research fails, phone calls and letters could produce unexpectedly productive results. Even when the initial recipient of a call or letter can't help, they often provide names of people who can and do help. Return kindnesses by offering to share what you've learned with your newfound cousins and friends. Don't be discouraged if you find little pertinent information about some women. Get to know each woman as intimately as possible. Women who left an unmarked trail remind us to leave detailed footprints ourselves. Start keeping a journal now.

Enjoy meeting my women and the search for your own women. Every woman's story is unique and worth our effort to unearth. We benefit from examining their triumphs and their failures. Patiently study women who appear to have failed to exemplify God's best. Because I too have often failed, those women give me hope. We can learn from their shortcomings, ask God's forgiveness, seek His help, and then improve. Let's focus on spiritually strong women and follow their lead.

Heaven is crowded with women cheering for us and speaking blessings on our efforts. Even before we learn their names and stories, we thank God for spiritually strong women who've come before us.

Selena

The Women and Their Relationships

Selena's mother

Grace

Selena's grandmothers

Lena (Grace's mother)

Marie (Grace's mother-in-law)

Selena's great-grandmothers

Annie Belle (Lena's mother)

Julia (Lena's mother-in-law)

Mamie (Marie's mother-in-law)

Daisy (Marie's mother)

Selena's great, great-grandmothers

Hannah (Julia's mother)

Tabitha (Annie Belle's mother)

Mariah (Annie Belle's mother-in-law)

Miriam (Julia's mother-in-law)

Cordelia (Mamie's mother)

Sarah (Daisy's mother)

Annie (Mamie's mother-in-law)

Therese (Daisy's mother-in-law)

PART I
Mother

Grace

1922 – 1982

When I began researching the women from whom I descend, I mistakenly thought researching and writing about my mother's life should be easy. I lived with my mother most of the first two decades of my life, and I knew her intimately for thirty-five years. For about thirty of those years, I was old enough to be aware of my mother's spiritual life (or at least to observe her worship habits). Yet I found writing about my mother a difficult, emotion-filled challenge. Although I knew the facts of my mother's life, facts did not sap my energy and rattle my peace of mind. Honest evaluation of a mother/daughter relationship can stir conflicting emotional memories: love and dislike, peace and regret, joy and pain, laughter and tears.

Long after I drafted Grace's chapter, I came back to it with a new idea. Before we research the lives of our mothers and attempt to document our findings, we should perform some helpful preliminaries. If you are a perfect daughter of a perfect mother, you may skip the next two paragraphs. If (like me) you're *not* perfect, please read on.

First, daughters need to confess their own shortcomings and be forgiven. If my mother were alive, I would ask her forgiveness for those times I was a difficult daughter. Since I studied Grace's life after she passed away, I need God's forgiveness for my failures in our relationship. He is faithful to forgive.

Second, daughters need to forgive mothers for their flaws. As God graciously forgives us, we must forgive our mothers. If your mother is living, make every effort to love her actively. If you can't love your mother under your own steam, ask God to love her through you. Honor your mother by being an honorable daughter. If you still need peace, ask God to help you find ways to reconcile differences and heal wounds. He is a worker of miracles. You might also need a great deal of God-provided patience. Simply making the decision to forgive will help clear your mind and heart to look honestly at your mother's life. Now I'll tell you about my mother, Grace. May my thoughts provide guidance and hope for your own journey.

Ideally, a mother is a person from whom we receive consistent, unconditional love. I believe my mother loved me to the best of her ability. I admired the positive character traits she exhibited, yet I disliked some of her lifestyle choices. Because Grace was talented at keeping up public appearances, people who knew Grace casually knew only the person she wished them to see. Including Grace's private struggles initially gave me pause. Because other daughters may face similar situations, I'll openly share all I know of Grace's story.

As you read the significant points of Grace's life, particularly those prior to my relationship with her, think of your own mother and decide how you might discover the chapters of your mother's story that occurred before your relationship with her. Viewing your mother's entire life and her interactions with other people can help you relate to your mother on an adult level if she is still living or help reconcile your memory of her if she has passed away. Now I'll follow my own advice and begin Grace's story from the beginning of her life. In the end, hers is a redemptive story of hope for us all.

Friday, January 6, 1922, dawned clear and nineteen degrees in Bonlee, a small closely knit community in the Piedmont area of North Carolina. During that cold snap, Grace was welcomed into a big, warm family consisting of her parents, Marvin and Lena, and six siblings ranging in age from fifteen to two. According to Mother Goose, Friday's child is loving and giving. With six children already under her wing at the time of Grace's birth and two more to come, Lena had no time for nursery rhymes. Grace probably did not read those words until she had given birth to a Friday's child of her own twenty-five years later. Mother Goose's prediction correctly applied to Grace. Her natural inclination was to be a loving, giving person.

Forenames fascinate me. My curiosity about a person's first and middle names usually prompts these questions: Was the person named for a family member? Did the child's parents like a particular name or its meaning? My mother's name, Grace Chadwick, honored a woman her parents admired. Prior my mother's birth, Mr. Chadwick and his wife, Grace, visited Lena and Marvin's home as members of a hunting club Marvin operated to help support his family. Mrs. Chadwick impressed Grace's parents by painting a portrait of one of Marvin's favorite hunting dogs.

Perhaps Lena and Marvin hoped their daughter would receive the gift of artistic talent. Although not expressed in precisely the same manner, Grace did become artistic. She became a hair and make-up artist. Grace never thought of herself as being graceful and occasionally laughed at herself for being clumsy. Most people called Grace by her first name. Her father called her Stewie, describing her always-bubbling (like stew) childhood energy. Several loving nieces and nephews referred to her as Gracie; to me, she was always Mama.

Grace's first head of hair was light blonde ringlets. Her hair turned light auburn by her twenties and, to her dismay, remained curly all her life. Her eyes were a pretty, medium blue under blonde brows and pale lashes. Grace's skin reflected a bit of golden color inherited from her father's Cherokee ancestors. She believed a suntan improved her coloring. Her mother, in an unkind tone, frequently reminded Grace her skin was sallow and criticized Grace for sunbathing. Grace ignored Lena's advice, but later regretted the premature aging that years of sun exposure caused.

Grace's dislike for her complexion and frustration with her too-curly hair prompted her career choice. She wanted to help other people by taming their hair and improving their skin with makeup. Despite Grace's dislike for her skin and hair, photographs prove Grace became a beautiful young woman although she said she never felt pretty. Lena's negative comments damaged Grace's self-esteem. Grace carried a feeling of inferiority all her life because of those few critical words from her mother. A mother's tongue can be a powerfully positive or negative instrument. May all women speak positive words.

On December 27, 1924, Grace's younger sister, Carol was born. Carol joined the family a few days before Grace's second birthday and became Grace's lifelong pal. The family's much-loved baby brother and last sibling, John Owen, was born in January 1927.

Eleven people living under one roof even in a moderately large home meant the children often slept two or three to a bed and ate their meals sitting on a crowded bench. Grace's older siblings occasionally had new clothing; she and the younger children wore hand-me-downs. The struggle to feed, house, and clothe eleven people drove Grace's father to work for a local lumber mill to supplement his income. Marvin continued to breed and train hunting dogs, lease land, and host hunting parties for wealthy New Yorkers and Pennsylvanians. He later left the lumber mill for a career as a baseball scout while continuing the hunting club to a limited degree. Meanwhile, Grace's mother occasionally clerked in a local grocery store. This book includes a chapter about Grace's mother, Lena, whose story sheds more light on the early years of struggle in the family home.

As Grace and her sisters grew older and were closer to the same size, hand-me-downs stopped flowing. Limited funds forced the girls to share their sparse wardrobes simultaneously. Grace and her sister, Evelyn, often repeated the fondly remembered *sandal story*. In the spring of 1935, Evelyn needed shoes for high school graduation and Grace needed shoes for eighth grade graduation. They scraped together enough money to purchase one pair of sandals. Grace wore the sandals for eighth-grade graduation in the morning (with socks because they were too large for Grace) then Evelyn wore the same sandals with stockings for her evening high school graduation.

Grace grew up in close proximity to several other large families in their rural neighborhood. Neighbor children who came to Grace's home were always welcome to stay for a meal. Somehow, there was always enough food. Neighbors welcomed Grace and her siblings to their tables as well. Grace cherished her childhood pals and maintained close friendships with many of them throughout her life.

Grace often spoke of happy school days. Of all her voiced recollections, my favorite was the story of how, in winter, she had walked to school with a hot sweet potato in her pocket to keep her hands warm. At lunchtime, she went to the cloakroom, retrieved the then-cold sweet potato from her coat pocket, and ate it from her hand as if it were an apple. She made it sound like manna from heaven. Early one morning when I was about four years old, I begged her to bake one for me. Soon the heavenly aroma of baking sweet potato wafted through our New York apartment. Grace called me to watch as she put the potato on the counter to cool until lunchtime.

At noon, I marched into the kitchen expecting a delicious treat. Grace handed me the cold sweet potato and I happily took a big bite then looked up at Grace with a frown. The potato Grace's mother had given her sounded good in the story but wasn't nearly as good to eat as the buttered sweet potato Grace usually gave me. In that moment, I realized for the first time that my mother was a person who had needs and I felt sorry for her. Years later, Grace laughed when she recalled my reaction. She didn't need my pity. Grace was grateful for a childhood crowded with loving parents, siblings, and friends. The sparse food and clothing of her youth made her an appreciative adult.

Grace didn't receive extensive Christian education led by her parents, who were busy meeting the physical needs of their growing family. Her exposure to matters of faith occurred when the family worshiped together and when she attended other activities offered by local churches. Grace's father was Quaker and her mother was Methodist. From the time of their marriage, however, Lena and Marvin worshiped primarily as Methodists. Although her family joined Hope Methodist Church in Bonlee, Grace also attended Bonlee Baptist Church during her childhood.

Thanks to foundational Christian education during her youth, Grace accepted Jesus Christ as her Savior and was baptized. Grace, her siblings, and their friends continued meeting their social needs and receiving spiritual food by participating in what was then called the Baptist Young People's Union. Grace later said her spiritual growth fell short in those early years because she failed to make the principles of her faith foremost in her lifestyle or allow them to inform her decision-making processes. Eventually, Grace did return to her spiritual foundation although not soon enough to avoid the negative results of unwise choices.

Grace loved music. Grace's maternal grandfather provided a second-hand piano for their family and although none of the children became accomplished pianists, they enjoyed hearing their mother and visitors play. Grace enjoyed singing along with the radio when she listened to popular music. She learned hymns at church and in the front room of their home, affectionately called the piano room. *In the Garden* was the first hymn she learned at home, enjoyed singing it in church, and it became her lifelong favorite. Grace loved to dance and was good at it (especially the jitterbug).

An athletic young child who played outdoors as much as possible, Grace excelled at girl's basketball in her teen years. Her childhood athleticism provided the physical preparation for golf and tennis as an adult. She and her childhood friends swam in nearby creeks and ponds where Grace developed the stoke she called the *dog paddle*. Her athletic interests also fostered her close bond with her father.

Every person in Grace's childhood family had to perform assigned chores. Grace had time for athletics because two of her older sisters excelled in cooking and housekeeping. Grace's less time-consuming chores included milking the cow (most dreaded on cold mornings) and killing the Sunday chicken when her father or brothers weren't home to take care of that unpleasant task. Her interests eventually grew toward a bit of decorating and cooking when she was a 1950s housewife, although homemaking was never her first love.

Dating was a group activity during Grace's teen years. Shortage of funds and limited transportation restricted the variety and frequency of social events. She and members of her social set were grateful when someone had access to a car for transportation to dances or the movie theatre in a nearby town. Grace lived in a dry county and didn't drink as a young person. She preferred to listen to music, dance, talk, and smoke. Beginning at the age of twelve, she smoked unfiltered cigarettes for the rest of her too-short life.

The death of family members can be a sobering event in a child's life. Her maternal grandparents passed away before Grace was born. Her paternal grandfather passed away when Grace was ten in 1932 and her paternal grandmother died when Grace was thirteen in 1935. Grace said she didn't have significant relationships with her paternal grandparents and didn't mention their deaths as major events in her life. Grace had spent little time with them because they lived about thirty-five miles away and transportation was limited. Grace thought of death as an expected part of human existence. She wasn't one to dwell on loss.

Perhaps Grace learned to cope with death while she was still living at home surrounded by a loving family. Grace's oldest sister, Mildred, left home to attend college, married, gave birth to a daughter, became ill, and died in 1936. Mildred's daughter, Barbara, died a year later. Grace often spoke of her beautiful little niece and chose to remember the gift of Barbara's brief presence in her life rather than dwelling on her death.

When Grace graduated from high school in 1939 at the age of seventeen, she wanted to turn her amateur hair-styling talent into a career. Her parents arranged for her to attend a cosmetology school in Norfolk, Virginia, where she could live with her mother's brother and his family. Grace was grateful to Aunt Marie, Uncle Ted, and their two daughters who welcomed her into their home and fed her well, although Grace complained of a chocolate pudding shortage. Pudding mix was a new treat to Grace and she thought chocolate pudding was the best food she had ever tasted. Her aunt divided the pudding made from one 4-serving box into five servings. As Grace scraped the bottom of her little pudding bowl, she dreamed of earning enough money to have all the chocolate pudding she wanted. She must have scratched that itch before my time because I never recall having chocolate pudding at my mother's table; however, Grace did continue to indulge in everything chocolate that came her way.

After graduating from cosmetology school, Grace returned to North Carolina and landed her first job in a town six miles from her parents' home. Ambition to further her career led Grace to a job in a larger salon about thirty-five miles away in the city of Greensboro. In October 1941, she packed her clothes and moved to live in a rooming house near her new job where her career continued to advance. She attributed her early success to the help of a kind woman named Vera, with whom she developed a lasting professional and personal association. Vera offered advice to help Grace please customers and develop her hairdressing skills. Perhaps because Vera helped Grace, my mother later helped other women succeed.

Grace worked hard and her satisfied customers tipped well. For the first time in her life, she had plenty of money. She had enough to pay her boarding house bill and to order whatever she wanted to eat at lunch counters. She carefully managed the rest of her funds to achieve her goal: to buy new clothing every week. To Grace, a complete outfit was a suit or dress and accessories such as shoes, a hat, gloves, and costume jewelry. Each Sunday, a pretty outfit donned, Grace went to services at the beautiful West Market Street Methodist Church.

In the spring of 1942, Helen and Godfrey, a newly wed couple living in Grace's rooming house, asked Grace to join them for a double date. Grace was to be the blind date of Godfrey's acquaintance, John, a young man who had recently arrived in Greensboro from Richmond, Virginia, to work as an announcer at the local radio station. Handsome, dark-haired, blue-eyed, smooth-talking John displayed gracious manners, and was impeccably dressed. Unlike anyone Grace had known, she thought John intelligent and sophisticated. Grace and John sat in Godfrey's car and talked until dawn. During several months of dating that followed, John met Grace's parents, her siblings, and her friends. Although not immediately apparent, John (an only child) was jealous of Grace's numerous family members and friends.

Grace's parents announced her engagement to John in mid-September 1942. A week before her wedding, Grace was entertained at a bridal shower where she and her friends enjoyed confectionary lovebirds while they played bridge and rummy. Grace purchased a wedding outfit: a *salute blue* dressmaker suit with a lighter blue crepe blouse, a veiled black hat with light blue trim, and black accessories. At three in the afternoon on Thursday, October 1, 1942, Grace and John entered West Market Street Methodist Church together, followed their two attendants down the aisle, and were married.

Although Grace's father disliked John, Marvin did not voice his concerns to Grace before her wedding. Marvin avoided giving his daughter away by abruptly leaving town to attend the baseball world series. Grace was understandably hurt. Marvin's concern was probably God prodding him toward fatherly intervention, yet Marvin failed to listen and refused to act. Grace's mother, who disliked John as well, attended the wedding and said nothing. Lena's policy was to avoid giving any advice to her adult children and she specifically refused to comment about their choice of spouse. Grace's parents should have attempted to share their views even if Grace ultimately ignored their concerns. Their parental radar was picking up appropriately apprehensive signals.

At the age of twenty, newlywed Grace had little life experience. She assumed her husband loved her and would treat her with kindness. She had watched her parents work together to weather life's problems and build a family. She had felt secure and loved by her parents, siblings, extended family, and community.

When John was eighteen months old, his father disappeared. His mother, a marvelous example of good character and diligence, taught school to support them; yet John had no father to model the role of a good husband. John's maternal grandparents modeled a good marriage when he lived in their home, yet John failed to follow their lead. Both of John's grandfathers were successful Methodist ministers, so Grace assumed John's extended family had provided a spiritual foundation. A perfect gentleman during their courtship, John became critical and abusive early in their marriage.

After their marriage, Grace continued to work as a cosmetologist and John worked as a radio announcer in Greensboro. In late 1944, John's radio career goals necessitated a move to New York City. Grace and John immediately loved New York where they found small-town-like associations even in such a densely populated city. Grace didn't need to rely solely on the kindness of strangers to feel at home in New York. Her older sister, Evelyn, had moved there with her husband, Hamp, a few years earlier to further Hamp's career as a musician. John and Grace rented an apartment adjacent to Evelyn and Hamp's apartment in Sunnyside, Queens. When Evelyn and Hamp brought their newborn daughter, Sally, home from the Boulevard Hospital in February 1945, Grace was thrilled to be living next door to help care for her niece. She grew to adore Sally more as a first daughter and Sally loved *Gracie* in return.

John arrived in New York with no job or contacts, a brief resume, and a limited professional reputation. He believed he could talk his way into a job. Publicly articulate and charming, he did quickly land a position as a news commentator with a national radio broadcasting system. His need to be a star in his field seemed at war with his feelings of inadequacy. John's searing verbal abuse of Grace worsened and became physical.

When John made an on-air mistake, he drank heavily. When drunk, he abused Grace. That routine became the norm. Emotionally crushed, Grace was also physically battered. Grace believed, as women in her situation typically do, that somehow John's dissatisfaction was her fault. Embarrassed, she told no one what was happening. The shouting and beatings did not remain secret for long, however, because Grace's sister and brother-in-law lived in close proximity.

Evelyn did not hide her feelings or hold her tongue. Evelyn's kitchen window, aligned across a narrow alley from Grace and John's kitchen window, gave Evelyn a front-row seat when John threw a glass of milk on Grace then slapped her face. Hamp had to restrain Evelyn who was determined to go through the windows to stop the assault. Evelyn wanted Grace to leave John immediately and strongly disliked him from that point on. Although Grace later realized she should have listened to her sister, she didn't leave and the situation grew worse. Jealous of Grace's love for her family and wanting to avoid interference from Evelyn, John began saving money to move to a different apartment in New York. John's drinking increased and in an effort to please her husband, Grace drank with him. Drinking turned out to be the worst habit either of them could develop.

Grace and John neglected their spiritual life. They attended church together in North Carolina but church was not part of their life in New York. John refused to attend church and did not want Grace attending alone.

Since John's chronic bronchitis kept him out of active service during World War II, Grace said the only wartime impact on their household was the shortage of goods. Limited funds and food rationing made grocery shopping difficult. They frequently enjoyed meals in a neighborhood Italian restaurant where the friendly proprietress served large portions for little money. By 1946, a typical American-style meal in their neighborhood included pork chops, vegetables, and rolls with butter for eighty-five cents. Nutrition was important by the autumn of 1946 because Grace was pregnant.

Grace and John's only child, I was born in June 1947. The myth that a bad marriage improves with the addition of a child proved false, unsurprisingly. Grace was a loving, caring mother. John worked long shifts, came home, drank, and abused Grace. Beatings and verbal abuse left Grace emotionally and physically damaged. She was once so badly beaten she crawled up the street to a doctor. Grace's parents occasionally argued but the concept a husband beating his wife was foreign to Grace. She believed that if she could somehow please John, the abuse would stop.

Grace continued close relationships with her parents and siblings. Two of her brothers served during World War II and returned home safely. Tragedy struck in May 1949 when Grace's older brother, Lynn, died accidentally in his home. Lynn's friends, family, military colleagues, and those who knew him as a teacher and coach admired him. Grace and her siblings described Lynn as having a sparkling personality. Although her brother's death was shocking to Grace, she repeatedly said the greatest tragedy was that Lynn's son lost the opportunity to have a long relationship with such a well-loved man.

Grace's focus in the late 1940s and early 1950s continued to be mothering me and being the gracious southern wife who made a good impression on John's business associates. She was an asset to him. John achieved his goal of moving Grace away from her sister when we moved to the Long Island neighborhood of Forest Hills in 1949. Although Grace missed being close to Evelyn and her family, she made friends easily and quickly became a typical 1950s housewife in a social circle of women who had young children. The mothers played bridge, shared lunch, and created group outings to take their children ice skating and shopping as well as to the playground, school events, and birthday parties.

My memories of Grace when I was young and we were together alone are sweet. It seemed to me there was never a dull moment. Even when we spent an entire day in our apartment, I was never a lonely child. She ate the cakes and cookies I produced in my little oven, complimented my finger paintings, and introduced me to literature. Before I could read, Grace read aloud to me; initiating for my lifelong love of books.

Our favorite author, E. B. White, produced such well-written children's books that Grace was entertained as well. We read and re-read *Stuart Little* and *Charlotte's Web*. A favorite short story, *The Little Match Girl* by Hans Christian Andersen, ended with a little girl's grandmother lifting her to heaven. When we read *The Little Match Girl*, we always cried at the happy ending. I learned tears could improve the drama. We agreed, however, that tears of joy at a happy ending were quite different from sad tears.

We laughed a great deal during my early years. Acting silly was something Grace fully sanctioned and actually taught. Some days she dubbed *operetta day*. No talking permitted. We sang every word. Other days were *pig Latin* days. My mother spoke English and had no command of other formal languages although she was expert at *pig Latin*. I struggled to understand her until she taught me to move the first letter or syllable of each word to the end of the word and add *ay* or *yay* sounds. Grace could make any day fun. Even during years of abuse, she created a happy life for her child.

Grace provided training for proper childhood behavior. By the time I was two years old, she expected me to dine in restaurants and behave myself so she conducted table manners practice at every meal. She reported no mishaps so she must have been a good teacher. New York restaurants made fancy children's versions of my parents' cocktails, provided amazingly delicious treats such as special desserts, and generally created an experience that taught me good behavior could be rewarding.

Well-mannered speech was important to Grace who instructed me to say *Mam* and *Sir* when addressing my elders. Grace was displeased when my first-grade teacher reprimanded me for saying *Mam* because, the teacher said, I was not a servant. After hearing my report of that reprimand, Grace met with my teacher and insisted the teacher support Grace's efforts to raise a respectful child by permitting me to address her as *Mam*.

Grace also taught rules governing my behavior at adult gatherings. We practiced introductions to adults during which Grace taught me to look directly into their eyes, smile, curtsy, and speak clearly. Following introductions, I was instructed to fall silent unless or until I was invited to speak, and to refrain from interrupting adult conversation. John expected good manners; Grace was the teacher, encourager, and disciplinarian (when necessary). I could be willful and stubborn yet I knew better than to misbehave in public.

To Grace's credit, she made lessons in decorum pleasant. She taught primarily through affirmation and smiles. When necessary, Grace did rightfully apply the *rod* by using her open hand on my clothed bottom. Her few spankings were, I believe, administered correctly; I remember no pain yet I was humiliated and contrite. As I grew older, she dispensed discipline in words and a *look*. Grace perfected the ability to transmit the direct, expression-filled mother's glance (or stare, if necessary) that powerfully stops the offending child mid-step without disturbing others in the vicinity.

Although New York City was a great place to live, Grace eventually became homesick for the country and close family ties. Beginning in the late 1940s, Grace and her siblings met every summer at their North Carolina family home. In addition to the social joy of family gatherings, big-city health scares (such as polio) prompted parents to take children away from cities and into the country. Adults wanted to give their children the clean, country summers they had enjoyed themselves. Those summers were a time of peace and pleasure. Grace enjoyed being with her brothers and sisters, their equally loved spouses, the growing group of nieces and nephews, and childhood friends who dropped by with their children.

Grace and her sisters organized memorable trips to clear, sandy-bottomed White Lake, where children safely romped while adults sat chatting nearby. Grace and her sisters rented a cottage where we slept in bunches on beds and on the floor. We ate meals on a screened-in porch. White Lake visits were simple, blissful escapes. Grace enjoyed the peace of being away from John who created new problems in New York during our absence.

John had affairs while Grace was away. After one summer vacation, John's female companion telephoned Grace after we returned to New York. The woman reportedly said she had lived in our apartment all summer and wanted Grace to divorce John so she could marry him. The thought of ending her decade-long abusive marriage might have sounded tempting to Grace; however, Grace responded that because she and John had a child, divorce was out of the question. Years of affairs and escalating abuse continued.

In 1954, John's employer asked him to move to Colorado to manage the news department of a Denver radio station. John went to Denver in late 1954 and Grace and I joined him in March 1955. At the age of seven, I was amazed to see my father's face pictured on billboards and in newspapers announcing his position at the local radio station. People stopped us on the street and came to our table in restaurants to meet him and welcome us to Denver. During one restaurant meal, John stood to shake hands and chat with a couple who had come to our table. They complimented John's on-air work and the improvements in the station programming; then he responded in his usual, articulate, handsome persona. I leaned over to my mother and whispered, "They don't know what he is like at home, do they Mama?" When I spoke those words, Grace realized I was keenly aware of the problems in our home. Although she had been willing to continue in her marital situation, she realized she needed to consider the safety and emotional well-being of her child.

Life grew progressively worse in our home. John's drinking was out of control. He had mismanaged his funds and could not pay moving and storage costs to retrieve the furnishings shipped from New York. We were living in a sparsely furnished basement efficiency apartment. John's emotional and physical abuse progressed. During a cookout in the summer of 1955, a neighbor's innocent, polite compliment about Grace's dress triggered John's jealous rage. He went inside, locked the door (I was already sleeping inside), and refused to let Grace into the apartment. She slept that night on top of the washers in the utility room adjacent to our apartment.

After John left for work the next morning, Grace awakened me by tapping on the window over my bed and asking that I let her in the front door. When I asked why she was dressed in the same clothes she had been wearing the night before, she told me she had slipped them on to do the laundry and had inadvertently locked herself out. By then, I was old enough to know she didn't do laundry in a party dress. Grace realized she wasn't fooling her child or our new neighbors. Although my father had never harmed me or even threatened to do so, Grace feared his new tactic of locking her out could cause John to move on to abusing me.

Later that day, Grace telephoned her sister, Evelyn, who was then living in Virginia. Delighted to hear Grace was finally leaving John, Evelyn arranged for us to travel under assumed names. Evelyn and Sally were to visit a friend in California so Evelyn routed their return via Denver and we were to join them on a flight to their home in Virginia. From Virginia, Evelyn offered to drive us to North Carolina. Beyond getting to the safe haven of her parents' North Carolina home, Grace had no idea what we would do. She was certain it was time to get out.

On the Friday morning before Labor Day 1955, Grace behaved normally as John prepared for work. Once he was out of sight, Grace quickly packed our suitcases, we put on our traveling clothes, Grace called a taxi, and we left for the airport. A kind neighbor stored, and later shipped, a trunk containing the rest of our most precious belongings.

Grace had endured all those years of abuse with no external spiritual support. Regular church attendance Grace enjoyed during John's courtship and the early years of their marriage stopped at John's insistence. She complied with John's wishes yet knew her child needed a spiritual foundation. She taught me to pray, taught me to sing hymns, and walked me to the nearest Protestant church for Sunday school and Bible school. Grace attended church when I had a role in a special program.

Although both of his grandfathers were Methodist ministers, John became emphatic that he was an atheist. Grace did not view his atheism as an expression of a well-thought-out decision; rather, she thought his professed atheism reflected his mistaken belief that atheism indicated intellectualism.

After fifteen years of marriage and more than a decade out of the workforce, emotionally drained Grace was back in her childhood home. She faced raising a child alone without funds or a job. Fortunately, Grace had loving parents who offered safety, housing, and food. So many women in similar circumstances have no family members to whom they can turn. Within a few months, Grace had renewed her state license to work as a cosmetologist and had secured a job. Before long, she and a friend opened a salon together where they worked diligently and enjoyed apparent success for several years but made little profit.

Varicose veins soon riddled Grace's legs and her hands peeled from chemicals. The shop opened early for women who worked early shifts at local mills and stayed open late for those who wanted to come after work. Occasionally, my mother and her co-workers went to the homes of long-time customers to dress their hair for special events such as weddings and proms. For many years, Grace accompanied winners of the Miss Siler City pageant (the local Miss America preliminary pageant) to the weeklong Miss North Carolina pageant. Grace served as the contestant's chaperone, manners consultant, make-up artist, and hairdresser.

Grace also began a merciful practice that gave me a glimpse of her love for others. When her regular customers passed away, their families sometimes asked Grace to use her hairdressing and makeup skills to help make their loved one look lifelike. Even though the local funeral home employed other cosmetologists, the relatives felt someone who had spent years styling their loved one's hair could better create the proper image of normalcy prior to the funeral home viewing. My mother seemed to take seeing the lifeless bodies of her long-time customers and friends in stride. I saw her weep only twice. Once, a young girl had died of leukemia and although Grace had frequently styled the girl's hair and had seen her health gradually decline, it was still shocking for one so young to pass away.

The other instance involved a troubled woman who had taken her own life by drowning. As a kindness to the grieving family, Grace agreed to style the woman's hair and apply her makeup. When Grace arrived at the funeral home, the staff had not had time to clean the woman's hands thoroughly. Grace never forgot seeing mud under her friend's fingernails. Grace believed she was seeing evidence of the woman's struggle to hold herself at the bottom of the pond.

Unfortunately, even though Grace and her friend were popular local hairdressers who worked long hours for a seemingly endless parade of customers, financial concerns eventually drove them to close their shop. Fortunately, the owner of a newly opened drug store asked Grace to manage the cosmetics department. Grace accepted the position and spent the rest of her career in cosmetics.

After leaving her marriage, Grace struggled to find appropriate living arrangements. After two years of living in her parents' home, she began to feel the need to establish a home in the larger town of Siler City near her job and where the school system offered subjects not available in the school near her parents' home. Grace's brother and sister-in-law, John Owen and Rachel, had established a home in Siler City then John Owen left to supervise a construction project out of state. Rachel invited us to share her duplex apartment near a good school until Grace could secure an apartment.

Beyond giving Grace peace of mind by offering adult supervision for me after school until Grace came home, Aunt Rachel's love provided healing for me. Rachel shared her home and cat, taught me to sew, and answered my ten-year-old questions about physical changes girls undergo as they become women. By the time we moved to our own apartment a year later, Grace thought I should have been Rachel's daughter because our interests were so similar. Grace seemed happy to share her daughter with Rachel, a sister-in-law she loved and admired for decades.

From 1958 until early 1963, Grace rented an apartment for us. It wasn't glamorous. Friends and family provided used furniture, we had great neighbors, and Grace purchased bunk beds so I could have friends over for pajama parties. Unfortunately, Grace's own sense of failure at marriage and stresses of work drove her to deem numerous evening drinks a necessary anesthetic and a way to forget life's unhappiness. She might have addressed her unhappiness more constructively, but she wasn't hearing positive input. Even though she made certain I attended Sunday school and church, she refused to go herself saying she was too tired or didn't have the proper clothes.

Grace had beautiful, classic clothes from the New York years. She believed in purchasing high-quality clothing and took good care of her wardrobe. Because she wore a uniform during her hairdressing years, she seldom wore her good clothes. When she worked in the drug store, she wore her own clothing protected by a smock. Grace did manage to have the right clothes to go out in the evenings, to play golf, and to dress for any occasion she deemed important. We all make choices. Sadly, Grace's choice to avoid church meant she was failing to hear encouraging words that might have made a difference in her outlook. She seemed to spend years stagnated by her sense of failure.

Grace's fifteen years of abuse diminished her self-esteem and triggered her alcoholism. Instead of seeking God's help, or choosing friends who spent their time in church activities or civic organizations, Grace chose friends who drank. Her unhappiness steadily deepened, her drinking increased, and her outlook on life became bleak. She lived in a defeated mode. Prior to her years of heavy drinking, Grace awoke each morning happy and ready for breakfast. She became a person who arose unpleasant bordering on angry.

It was painful for people who loved Grace to watch her unhappy, downward spiral. I hope women reading her story in the twenty-first century will seek the many resources that weren't readily available to Grace. Learning of Grace's struggles should prompt women who are suffering or who have suffered debilitating spousal abuse to seek help. Perhaps women who have succumbed to the use of drugs or alcohol in an attempt to numb their pain will realize the futility of those addictions.

Even when Grace seemed to turn away from God, He never turned away from her. It was evident God was watching over our home on one evening in particular. Grace fell asleep on our sofa from work exhaustion coupled with too much to drink. She dropped a burning cigarette on the synthetic upholstery and it began to burn. Grace awakened in time to put out the fire that could have claimed her life as well as mine as I slept upstairs.

In March 1963, events in our extended family caused another shift in living quarters and add stress to Grace's life. Grace's father suffered a fatal heart attack; two weeks later, her widowed sister died in a tragic automobile accident. Because Grace had no spouse, her siblings elected her to move back to her parents' home to help her mother and orphaned nephew. Feeling forced to give up her independent living space seemed to drive Grace further into a dark, unhappy place. Grace took up residence in her bedroom in the back of the house where she shut herself away most evenings. She confined her drinking to her bedroom and bathroom; so although I saw the results, I didn't have to watch the consumption first-hand any longer.

Although Grace wasn't happy about returning to her mother's household, it turned out to be the best move for her fifteen-year-old daughter. Initially upset that I could no longer walk to visit to my high school chums, I soon found peace and normalcy in a house where my grandmother kept a regular routine of rising at seven in the morning, serving three meals a day, then going to bed at eleven in the evening to read her Bible. The move placed me much closer to my church where, in addition to Sunday church activities, I could join Girls Auxiliary and later Young Women's Auxiliary led by godly women and attended by girls from Christian homes. When mothers are struggling, God can bless their children with other women as role models and teachers of positive character traits.

After John, Grace was wary of men. Between 1955 and 1965, Grace focused on making a living and parenting me yet she found time for a limited social life. Eventually, she met a divorced man with whom she shared interests such as golf, bridge, and dancing. They dated regularly for many years. He was always kind to me and I grew fond of him. He eventually proposed marriage; however, Grace was unable to trust anyone enough to marry again. A close family friend from Baltimore who had worked with her father in baseball also proposed marriage in 1965. Grace turned him down as well because she thought of him more as a brother. Aside from those two men, Grace went on a few casual double dates with friends.

The monotony of Grace's work life and sharing a home with her mother further accelerated her alcoholism. By 1972, Grace's evening drinking bled into her days and became an issue that threatened her job. After years of excessive alcohol use, she recognized the seriousness of her drinking habit when hospitalized for a health crisis. Grace's new life began in 1973. Although she did drink occasionally over the next nine years, she never returned to the alcohol-controlled life she had been living. Praise God.

Fortunately, when Grace recovered physically, she received a job offer to work in another drug store cosmetics department with people she greatly admired. It was the last position of her career and a happy time. Throughout her work life, Grace enjoyed knowing people of all social and racial groups. Customers, drawn to Grace's friendly manner, shared their happy and sad stories. In return for her concern and friendship, they shared their possessions. People gave her amazing boxes and sacks of seasonal, homegrown vegetables and fruit (they knew her favorites were tomatoes, corn, and dewberries), home-canned goods, cakes, pies, biscuits, cookies, and (her favorite indulgence) chocolate fudge.

The progress Grace continued to make in curtailing her drinking and being more positive was due, I believe, to a return to her spiritual roots. Once sober, she began attending church and seeking Biblical solutions to life rather than unsuccessfully attempting to mask problems with alcohol. Even though she didn't seek help from a formal alcoholism treatment program, she wasn't alone. God was holding her up and carrying her through. When Grace returned to Hope Methodist Church, she found several of her childhood friends, unconditional love, and no judgment. Even though her church's membership had dwindled to a few members and held worship services only twice a month, she participated in an adult Sunday school class regularly and even taught a few times. She once telephoned my husband and me to say that teaching was initially a frightening prospect. Once involved in Bible study, she found the process interesting and fulfilling. Hindsight is always so clear. She began to realize the wisdom she was gaining from her study could have helped her avoid earlier struggles. Grace's return to faith began to restore her peace.

Grace's living arrangements changed once more. When their mother's health declined, Grace and her siblings helped Lena move to a nursing home where she passed away in 1977. Grace moved into an apartment closer to her workplace and Carol, her widowed younger sister, moved in next door. It was the happiest time of Grace's life. She was working with people she loved, participating in her church, and spending evenings talking with her sister. The only unhappiness that interrupted their fun was when Carol and Grace's baby brother, John Owen, had a heart attack and died suddenly on October 28, 1981.

In July 1982, my husband graciously offered to treat Grace to a marvelous vacation. We invited Grace and her sister, Carol, to join us on a visit to Colonial Williamsburg in Virginia. Limited funds had meant Grace had taken few vacations. The trip was wonderful. When we returned home, Grace called to thank us for what she described as the best vacation she ever had and said she was certain she'd never experience another vacation as wonderful. She didn't know how prophetic those words would be. A few weeks later, Grace drove to church in her car then left by ambulance. She had a heart attack during Sunday school.

During the days following Grace's heart attack, Aunt Carol repeatedly told me there was nothing I could do and if Grace could speak, she would say she didn't want me to see her in her deteriorating condition. Thanks to Aunt Carol, my lasting memory of Grace was seeing her smiling on a Virginia vacation rather than lying in a hospital bed with her face swollen and distorted beyond recognition. Grace repeatedly said it was her wish to die quickly. God granted Grace's wish. She never regained consciousness and died at the age of sixty on August 25, 1982.

Initially, it was difficult to understand Grace's sudden death occurring when she finally seemed to be enjoying life. In time, I realized the Lord healed Grace's spiritual condition yet there could be no instant reversal of damage to her physical condition caused by decades of smoking and drinking. It was a relief to know she was finally at peace.

Three days after Grace's death, I sat under a canopy in the cemetery of her church near the graves of her parents and several of her siblings. Against a backdrop of rolling, green hills beyond that little cemetery, her pastor conducted the graveside service then closed his Bible. He ended the service by singing an a cappella solo of *In the Garden*, the hymn Grace had loved since childhood. It was a hot August afternoon yet as the pastor sang, a surprisingly refreshing breeze blew over us.

Grace would have loved that moment of peace and beauty. As Grace's favorite hymn states, we all come into God's garden alone. If Jesus Christ has been part of our lives through our acceptance of Him as Savior and Lord, He stands with us before God in eternity. As Grace's pastor sang, Grace was experiencing what her favorite hymn describes. Jesus was walking with Grace, talking with her, and telling her that she belonged to Him for eternity. I'm so grateful Grace's words and actions during her last years assure me she had freshly embraced her personal relationship with Jesus. She delayed but she came back in time.

More than a quarter century after her death, people say they remember Grace as someone who listened to everyone, genuinely cared about others, and was especially kind to older people and children. Through it all, Grace tried to love the people in her life. Sadly, she seemed unable to love or forgive herself for what she saw as her failure to have a successful marriage.

Through Grace's apparently effortless love for her family, friends, and acquaintances, she set a high example I might never attain but I'll work to improve. She tried diligently to avoid hurting people's feelings and treated everyone equally. She looked for goodness in every person and so should I. "Treat everyone you meet as you'd like to be treated," Grace advised. "Be kind and considerate," was her most often-repeated phrase. Grace was correct. Even when we follow Biblical teachings about sharing our faith with others, we should speak respectfully. When the Holy Spirit calls people to Him, we want them to see spiritual women as approachable, willing guides.

Relating to one's mother not only as a daughter but also as sister through mutual salvation is a double blessing. For thirty-five brief years, Grace was my mother. For eternity, thanks to God's amazing grace, this woman called Grace is an even closer relation: she is my sister in the Lord.

Grace was human so Grace was, by nature, imperfect. Because she was God's choice for the job, Grace was the perfect mother for me.

PART II
Grandmothers

Lena

1887 - 1977

If the phrase *she never caught a break* applies to any woman in my ancestry, it applies to Lena. Beginning in childhood, Lena endured loss of loved ones, hard work, and disappointments. I thought of Lena as second mother rather than as my grandmother. My personal relationship with Lena began when she came to New York to help feed and diaper me as a newborn, then I visited her home at least annually beginning when I was a toddler. I lived with her for two years when I was eight to ten years of age, then visited her on weekends until I was fifteen. Lena welcomed me back into her home as my permanent residence when I was fifteen for several more years. She taught me to cook and sew, the correct way to sweep a porch, and the proper procedure for making a bed. She allowed me to observe her regular Bible study. Knowing Lena positively altered my life. Here's her story.

Named Emma Lena at birth, Lena was born on Christmas Day 1887 in the home of Bob, a country dentist, and his wife Annie Belle. Their home was in Pleasant Garden, Guilford County, North Carolina. You'll read about Bob's family in the chapter of this book dedicated to his mother, Mariah. Annie Belle's story appears in her own chapter. We'll begin Lena's story on the day of her birth.

Lena was the second child born to her parents. Lena's mother had given birth to a daughter in 1886 who was stillborn or died on the day of her birth. Lena was the only child in the household for almost three years. Her brother, Banks, was born in 1890, followed by Ted in 1892, then Eula Belle, in 1896.

Lena's first name honored her mother's older sister, Emma, who died before Lena was born. Lena was the shortened form of a name her father had chosen. Bob had wanted to name their daughter Selena, the name of a girl he had courted before he married Annie Belle. The couple compromised by calling their daughter Lena.

When Lena was born, she had only one living grandparent. Her father, the seventh child of his father's second marriage, was born when his father was sixty-two years old and his mother was thirty-eight. They had both passed away before Lena was born. Her mother's father had passed away in 1865 leaving Annie Belle's mother, Tabitha, as the only grandparent Lena knew. Tabitha, whose story appears in her own chapter of this book, played a significant role in Lena's life. Lena's shortage of grandparents was offset by her many aunts, uncles, and cousins who lived nearby.

Lena's childhood church was Pleasant Garden Methodist Church. Because both her father and mother had relatives who worshiped at nearby Bethlehem Methodist Church and Tabernacle Methodist Church, Lena enjoyed revivals, singing classes and camp meetings at those churches as well. Until 1944, one pastor serving a circuit of churches pastored Lena's church among several. Although each congregation offered Sunday school weekly, they might not hold a worship service every week. If members wanted to hear preaching weekly, they could travel to a neighboring church for worship. Lena became a Christian as a young child and never doubted her salvation.

God knew Lena needed faith and strength early in her life. Her mother died of tuberculosis at the age of thirty-nine when Lena was only nine. Lena's maternal grandmother, Tabitha, left her nearby farm in the hands of her son, and moved in with her son-in-law. For three years, Tabitha mothered the children, kept house, trained Lena to perform homemaking chores, and provided emotional support during the little family's difficult time.

Three decades earlier, Tabitha had received support from her parents, Alexander and Rebecca, when her husband died leaving her with five children. Rebecca stepped in to help Tabitha so it probably seemed natural for Tabitha to step in to help Bob. Bob must have been grateful and I know from Lena's first-hand account of those years, the children felt loved and comforted by their grandmother, whom they grew to love as a mother.

In 1900, when Lena was twelve years old, Grandmother Tabitha moved back to her own home leaving Lena to take over housekeeping and mothering duties for the family. Lena must have felt overwhelmed; however, keeping house for her father and siblings was only a taste of the work Lena faced during her lifetime. Lena's little sister, Eula Belle, died of tuberculosis at the age of eight in 1904.

Because Tabitha was helping with housekeeping, Lena had time to attend Piney Grove School. She went on to attend the Academy (the equivalent of high school) while keeping house for her father and siblings. Lena said she loved learning to keep house and enjoyed all her subjects in school. Although photographs of Lena between the ages of nine and thirty-nine have not been found due to fires at her father's home and later her own home, my mother loved telling a story that provides a glimpse of Lena's beauty in young womanhood. An elderly gentleman, upon meeting my mother, said, "I can tell you're Lena's daughter but you'll never see the day you are as beautiful as your mother was."

Lena's father, having attended medical school two states away, encouraged her to use her quick mind to have a career before she married. She attended the first course of nursing school at St. Leo's Hospital in Greensboro although she didn't finish her training because, as she later said, "I took a notion to marry Vernon," a handsome farmer who lived nearby.

When I began to study Lena's ancestors and those of her first husband, I was amused to learn Lena and Vernon were distantly related. Their respective family trees are dotted with numerous intermarriages among Lena's people and Vernon's. Were they aware of their distant blood and marriage relationships? Lena's grandmother, Tabitha, might have explained that her own grandmother, another Tabitha, was from the same family as Vernon's father. If she did mention their kinship, their connection was too distant to stop the wedding.

Vernon and Lena were married in the parsonage of Pleasant Garden Methodist Church by their pastor, Rev. John Stover, on September 20, 1905. On their wedding day, Vernon was twenty-four and Lena was eighteen years old. An experienced homemaker, Lena moved her belongings to Vernon's small farm near her father's home, and enjoyed establishing her own home. Within one month of her wedding, Lena was pregnant. Grief soon overshadowed the joy of expecting her first child. Vernon became ill with diphtheria.

Called the *strangling angel of children*, diphtheria caused horrible suffering that frequently ended in death. Diphtheria's highly contagious pathology usually made quarantining the household necessary. Before inoculations for prevention and antibiotics for treatment were developed and widely available, 17,000 American people succumbed annually to diphtheria. Even little Archie Roosevelt, a son of then President Theodore Roosevelt, endured a nine-day struggle with diphtheria and temporary heart failure. Archie managed to survive. Vernon wasn't so fortunate.

Miraculously, pregnant Lena remained healthy. I don't know whether she helped nurse her husband or if someone else cared for him. Fresh from nursing school, perhaps she had learned to use caution to avoid becoming infected. Initially, Vernon probably had a sore throat and a low-grade fever. Diphtheria then creates a gray or white coating on the tonsils, the roof of the mouth, and the sides of the throat. Swelling in the neck makes swallowing difficult. Finally, the patient slips into a coma and dies. Vernon died on April 20, 1906, followed by burial near Lena's mother in the cemetery of Pleasant Garden Methodist Church.

Imagine a young bride watching her husband suffer. When Vernon died, Lena was six months pregnant. Without a husband to work the farm, Lena moved back to her father's home. Her father was then fifty-one years old and her brothers, Banks and Ted, were still living at home. Lena's son, Robert Vernon, was born on July 22, 1906, and named for his father and Lena's father. When asked about her early adult life, Lena replied, "I was a bride, a widow, and a mother in one year." With a young baby to care for, returning to nursing school was impossible. Lena taught pre-school children in her father's home for several years.

Lena often spoke of the close friendship she had with Kate, the wife of her cousin, Charlie, who lived next door to Lena's father during Lena's early widowhood and motherhood. Kate and Lena had much in common. They married the same year and gave birth to their first child a month apart. Lena also lived close to her church where she continued to worship and began the religious training of her son.

On April 17, 1910, the census taker recorded Lena's father's household as being headed by Bob, a widower (fifty-five) working as a rural route mail carrier (the only person employed in the household), sons Banks and Ted (then nineteen and seventeen, respectively), daughter Lena (twenty-two), and grandson Robert (three). Since the previous census, Bob's career changed because he lost his dental tools when their home burned. Bob couldn't afford to replace his tools so he secured a position delivering mail by horse-drawn carriage. Before the next census in 1920, the family changed radically.

For five years, Lena kept house for her father, brothers, and son. As her heartache over the loss of Vernon began to heal, a neighbor came courting. Marvin was a Quaker man who had been away to attend college and to play baseball. Marvin reportedly thought Lena was beautiful and began to court her.

Having known my dapper grandfather, I can imagine how he looked dressed for courting. Perhaps he appeared at Lena's door wearing a well-pressed shirt and tie with a fashionable hat in his hand. Marvin had received an excellent Quaker education including four years of college, was well read, and a challenging conversationalist. Marvin had also worked as a college baseball coach prior to their marriage. After five years of widowhood, Lena was ready for a new husband.

In the home of her best friend, Kate, and Cousin Charlie, Lena wed Marvin on October 23, 1911. Lena's Methodist minister, Rev. J. A. Sharp, officiated. Although Lena attended special events at Marvin's Quaker church, she wanted to continue worshiping as a Methodist, and Marvin was willing to change his denomination. They worshiped at Pleasant Garden Methodist Church with Lena's father and her son from her first marriage. They lived with Lena's father initially then with Marvin's family for a short time.

Lena loved Marvin's family. Marvin's parents, John and Julia, were both living at the time Lena joined their family. Marvin was the eldest of four children followed by Miriam (called "Mamie") who was then thirty and single (she never married). His sister, Lillian, then twenty-eight, was living in New York with her husband and daughter. His youngest sister, Mozelle, then twenty-six, had married in 1900 and was living in her own home.

Apparently more focused on other pursuits, the women living Marvin's family home (Julia and Mamie) didn't keep house to suit Lena. Lena was displeased that Marvin's family home had no window screens. They soon moved back to live with Lena's father. Marvin's youngest sister, Mozelle, lived nearby and was a dynamo of homemaking. Mozelle and her husband, Carl, had three children when Lena married Marvin. Mozelle's fourth child was a boy who became Lena's favorite nephew, Charles. Lena respected and loved Mozelle and the two women seemed to enjoy each other's company.

While still living in her father's home, Lena and Marvin welcomed sons Louis in 1912, and Lynn in 1913, followed by their first daughter, Mildred, in 1915. In 1916, Marvin moved the family to Bonlee, North Carolina, to a home large enough to double as a hunting club. Lena set up housekeeping in their first home, later called the Councilman house. Her father sent furniture and a piano to help furnish their new home. Marvin leased land, bred and trained hunting dogs, and organized parties of wealthy hunters. Lena gave birth to another daughter, Evelyn, in April 1917.

The family joined Hope Methodist Church. When their church didn't have worship, Lena and Marvin took their family to the Bonlee Baptist Church. As mentioned in Grace's chapter, the children enjoyed participating in activities such as the Baptist Young People's Union.

Lena's brother, Banks, was playing semi-professional baseball in 1917 when he developed tuberculosis. In January 1918, desperate to avoid losing another family member, Lena's father sent Banks to El Paso, Texas, for treatment at Sunny Crest Sanatorium. Lena never saw her brother again. Banks died in El Paso on March 19, 1918. No family members or friends were present to mourn his passing or attend his burial. Lena often mentioned her regret at not having been with Banks during his last days of life. She longed to visit his grave, which I have since done in her stead.

After Banks died, Lena's remaining sibling, her brother Ted, moved to Norfolk, Virginia, where he married, raised two daughters, and outlived her. She and Ted visited each other as often as their lives in different states permitted.

The 1920 census includes a record of Lena and Marvin's household in Bonlee. Recorded on January 7, 1920, the census noted Marvin's employment as a handle maker at a local lumber company (to supplement their income while establishing his hunting club business). The family members he was supporting in 1920 included a sixth child, Marie, who was born in May 1919. Lena's father, Robert (then sixty-six, retired, and in declining health), was also living with Lena and Marvin in 1920. At Robert's death in Lena's home seven months later, she lost a father who had demonstrated his love consistently and on whom had she depended for three decades.

Lena participated in Marvin's hunt club business by hosting wealthy hunters in her home. Lena's growing family kept her occupied all year, and she was overwhelmingly busy in autumn when groups of hunters arrived. During hunting season, she hired a local woman who arrived early every morning to help cook, clean, serve, and dress game. Lena and her helper cooked and served a big country breakfast then, while the hunters were eating, the two women packed a lunch basket for the hunters to eat in the field. Marvin organized hunting dogs, gear, and transportation, then hosted the day of hunting.

The hunters returned to Lena's home in late afternoon. They bathed, rested, met in the parlor for cocktails, and then moved to the dining room for a home-cooked meal from Lena's kitchen. A portion of the game killed during the hunt became part of evening meals; the rest was dressed and packed for the hunters to take home. Hunting parties usually stayed for at least a week.

To better accommodate their hunting club and growing family, Lena and Marvin moved a short distance down the road to a larger home in 1921. My mother, Grace, was born in their new house in January 1922. Lena gave birth to her eighth child, Carol, in December 1924. Her ninth and last child, John Owen, was born in January 1927. Lena's household headcount had grown to eleven. She and Marvin were busy providing sleeping accommodations, clothing, and three meals a day.

Lena made seventy-two biscuits for breakfast every morning to serve with dozens of eggs laid by their own hens, huge platters of pork chops, ham or sausage, and bowls of gravy. Commercial cold cereal was a rare treat enjoyed only on an occasional Sunday. Milk was fresh from their cow. Fortunately, Marvin, Lena and the children were healthy, and the children experienced only the usual childhood illnesses. Lena's ninth child, John Owen, experienced a problem with nursing when Lena's breast milk supply dwindled. Lena faced a further challenge when John Owen was allergic to cow's milk. They acquired a goat, John Owen tolerated goat's milk well, and then he began to thrive.

Finances and love of baseball drove Marvin to expand his work life. In 1929, he became a scout for professional baseball. He first scouted for the Cardinals, then for the Giants, and finally for the Brooklyn (later Los Angeles) Dodgers, the organization with whom he spent the remainder of his scouting career until he was in his eighties. Every spring, he visited North Carolina college teams. In summer, he watched semi-pro teams. Marvin's love of working outdoors, and his keen eye for identifying promising baseball players, made him well suited for his career as a baseball scout. For Lena, there were pitfalls.

Marvin's time away from home left Lena, the workaholic, to manage the large family without him. Marvin's preference was to talk with people or spend time with animals. The hunting dogs he raised were beneficiaries of his diligence. When he had many dogs to feed, he cooked big vats of food for them over outdoor fires. He bred dogs, raised litters, and trained them for the field.

After his baseball-scouting career began and hunting club activities diminished, Marvin's time at home was more leisurely. He spent more time sitting on the porch feeding squirrels by hand. He joked that he went indoors when Lena was mowing their large yard to avoid having neighbors talk. She didn't seem to mind doing the bulk of the work in their house and yard. It was Marvin's inclination to stray that hurt her deeply.

Marvin had an extra-marital affair with a woman he met during his baseball travels. Attempting to hide his continued correspondence with the woman, he rented a post office box in a neighboring town. The postmistress, Lena's friend, gave Lena his mail. Her friend's awareness of Marvin's deceit amplified Lena's humiliation at her husband's indiscretion. Although Lena stayed in the marriage, there was lingering bitterness between Lena and Marvin.

How did Lena endure? Leaving Marvin in charge of the Bonlee household, Lena spiritually and socially refreshed herself by attending annual camp meetings at Bethlehem Methodist Church near her former home. Each summer, she wore a pretty dress and broad-brimmed hat for the thirty-five-mile train trip to Pleasant Garden. Lena cherished visits with cousins and friends during the weeklong meetings.

Lena's youngest child was eight when her eldest daughter, Mildred, married in 1934. Mildred's family loved and welcomed her husband, John. On June 1, 1935, Mildred gave birth to Lena's first grandchild. Little Barbara was said to be a beautiful, good-natured baby. In October 1935, Mildred became ill with a thyroid condition described as an exophthalmic goiter. After treatment began, Mildred became pregnant again. When Mildred's illness necessitated hospitalization in May 1936, Lena stayed with Mildred day and night while Mildred's husband visited as often as possible. John worked to support Barbara and pay his wife's medical expenses. During Mildred's illness, baby Barbara lived in Lena's household in the care of Mildred's siblings who delighted in their little niece's presence.

Mildred became gravely ill. Attempting to save her life, the doctors terminated Mildred's pregnancy on June 1, 1936 (ironically, Barbara's first birthday). Mildred developed septicemia following the termination procedure. Mildred died at 4:30 a.m. on June 18, 1936. Three days before Mildred's twenty-first birthday, Lena was planning her daughter's funeral.

John graciously allowed Mildred's family to determine her burial location. Lena and Marvin's family graves were located in their hometown, and they expected to visit Pleasant Garden even if they remained in Bonlee. Mildred's burial took place on June 19, 1936, in the Pleasant Garden Methodist Church cemetery. Marvin, whose family burials were in the Quaker cemetery, allowed Lena to bury their Methodist child among the graves of Lena's parents, siblings, paternal grandparents, and dozens of relatives.

Mildred's widower, John, responsibly arranged for family assistance with daycare, and moved his daughter to live with him near his job. Accustomed to living with her mother's family, Barbara wailed when her father took her from Lena's house. Lena and her family heard Barbara's loud cries all the way to the end of their road as her father drove away. Sadly, Barbara developed whooping cough and died of a cerebral hemorrhage on July 10, 1937, at the age of two. Lena had lost her eldest daughter and first grandchild. She briefly honored the dead, remembered them always, but quickly returned her attention to meeting the survivors' needs. Warmly welcomed, John continued to visit Lena even after he remarried.

When deciding on the location of Mildred's burial, Lena and her family didn't anticipate burying Barbara. On July 11, 1937, Barbara's burial took place next to Mildred's grave; the spot later marked with a stone carved with a lamb and flowers. Mildred and Barbara were the only descendants of Lena and Marvin buried in Pleasant Garden. Visiting the graves often, Lena found comfort in seeing the beautiful sculpture of a mother holding a baby (donated by an artistic friend) situated between the headstones of her daughter and granddaughter.

Lena had lost her first grandchild; then God doubly blessed her on February 12, 1945, when she gained two grandchildren in one day. First, Lena and Marvin learned of the birth of Evelyn's daughter, Sally Kay, in New York. A few hours later, their son, Lynn, called from South Carolina to say his wife, Dottie, had given birth to a son. Although I'm not certain how quickly Lena saw Lynn's son, Terry Lynn, we have photographs of Lena holding Sally during the first days of Sally's life. Not long after receiving the news of her granddaughter's birth, Lena sped to New York by train to help with Sally while Evelyn recovered from childbirth.

Lena continued to welcome grandchildren. Her only namesake, I was born in June 1947, and Lena joined us in New York a few days after my birth. When I was two weeks old, Lena penned a letter to my other grandmother, Marie, who gave the letter to me decades later. Lena opened the lengthy letter, "Well, here I am playing nurse to our darling grandbaby." Lena informed Marie she's been sewing housedresses for my mother, caring for the new baby, keeping house, and cooking. When Lena's daughter, Marie, gave birth to a son in January 1948, Lena headed to Pennsylvania. Marie pleased Lena by naming her son Robert, honoring Marie's eldest brother, whose name originated with Lena's father.

Two of Lena's older sons, Louis and Lynn, were unharmed during their service in World War II. In 1949, Lena was unprepared when Lynn died accidentally. He died in his South Carolina home with his wife and four-year-old son, Terry, nearby. Accidental death rocks even the strongest families, and Lynn's death difficult for Lena and their family to bear. Lena internalized her personal grief, kept Lynn's memory alive by speaking of his many positive character traits, and focused her energy toward her daughter-in-law and grandson.

Louis and his wife, Dorothy, had three sons: Pat in 1949, Hugh in 1952, and Bill in 1960. Carol and her husband, Jimmy, welcomed two daughters and a son: Julia in 1952, Lisa in 1955, and Bill in 1957. Although she lost Barbara, Lena had ten grandchildren who survive in 2011.

Lena welcomed her growing extended family whenever they were able to visit. As I mentioned in Grace's story, Lena's children brought her grandchildren to the country to escape city diseases, and so siblings and cousins could enjoy long summer visits. Lena, of course, was in charge of the kitchen. She made delicious cornbread. Lena announced mealtime by removing big sheet pans of sizzling cornbread from the oven. Using the edge of her hand, she broke each pan of bread in half then quarters, stacked the bread on platters, and passed the platters to her grateful guests who then broke off individual pieces. She served her cornbread with cabbage, pinto beans topped with onions, green beans cooked with corn, peas, lima beans, turnip greens, yellow squash, or other fresh vegetables. The vegetables changed, yet the wonderful cornbread was a crowd pleaser and constant repast at Lena's table. If there was cornbread left over, it was crumbled into a glass and eaten soaked in buttermilk. Marvin's lucky dogs devoured cornbread dropped by toddlers or otherwise rendered inedible by humans. Lena's cornbread was never wasted.

Lena *said* she did not cater to the eating whims of individuals who dined in her home. She claimed to cook what she wanted to cook and said people could eat it or leave it. We knew better. She loved pleasing us with favorite foods. When Evelyn's husband, Hamp, came to visit, Lena made green beans with potatoes on top. She never made green beans that way unless Hamp was visiting. When Marie's husband, John, said he liked peanut butter on Lena's cornbread, she began adding the peanut butter jar to his table setting. When John Carey (Marvin's diabetic Dodger scouting partner) visited, he found containers of unsweetened strawberries and peaches in Lena's freezer marked with his name. He knew she had frozen the fruit especially for him and was grateful. To avoid Lena's wrath, her grandchildren interpreted the label bearing John's name to mean *do not eat unless you are John Carey.*

Non-human animals also enjoyed the benefits of Lena's ability to feed and nurture. Marvin usually provided the care and feeding to raise and maintain his hunting dogs. In at least one instance, he was wise to enlist Lena's help. I was living with Lena and Marvin in the winter of 1955 when Marvin's best Pointer gave birth to a litter of thirteen puppies. When the puppies were a few days old, the mother ate rat poisoning and died. Lena stepped in. She warmed milk, poured it into clean soda bottles topped with baby bottle nipples, and invited me to join her in a special project. Lena allowed me to help her feed the puppies their first meal of the day, then put them back under the heat lamp where they slept until she fed them at noon while I was at school. We fed them together again when I came home from school and again before bedtime.

Although Lena never admitted loving those puppies, I was a witness to her tender care, and the way she smiled with satisfaction as they hungrily consumed her offering. If asked, she said she took care of those puppies because they were valuable hunting dogs Marvin sold to supplement their income. Lena was pleased all thirteen puppies survived, and Marvin realized a substantial profit. Secretly, I knew Lena enjoyed the process of holding and feeding those adorable puppies. Underneath her tough exterior, Lena had a soft heart.

When my grandfather found an abandoned baby squirrel, Lena initially allowed us to keep him in her warm kitchen. She helped me feed Billy Herman, as we called him, with one of my doll bottles. Again, thanks to Lena's touch, Billy Herman grew to be a healthy, handsome squirrel and house pet until the day Lena banished him to the yard for pulling stuffing from her sofa. My rabbits, Bach and Beethoven, lived outdoors in a pen. Thanks to Lena, they enjoyed frequent indoor visits and feasted on cabbage, fruit peels, and vegetable trimmings.

When I brought home twin black kittens, Robert "Bob" Browning and Elizabeth "Beth" Barrett Browning, Lena fed them choice scraps, but was cool toward them in my presence. When I returned home earlier than expected, I found Lena napping in her favorite chair with Bob and Beth curled up in her lap. Lena even tolerated my hamster and, when he died on the day of President Kennedy's funeral, attended Hammy's somewhat less formal funeral in our yard.

Lena's interest in animals didn't stop with hunting dogs and pets. She loved wild birds and regularly scattered breadcrumbs for them in winter. To please Lena, Marvin purchased a birdhouse appropriate for Lena's favorite songbird, the eastern bluebird, and mounted the house on a post across the road from their home. Bluebirds found the house where, to Lena's delight, they nested annually. During my childhood, I recall Lena's stern warning to her grandchildren. Lena told us we were welcome to watch the birds from a distance, but strictly forbidden from going near the house or disturbing the inhabitants of Lena's Bluebird Inn. She was kind to God's creatures.

Whether Lena was feeding people or puppies, sweeping the porch or making curtains, mowing the lawn or painting furniture, feeding one infant or a group of hunters, her approach was consistent. She assessed each task, gathered raw materials and tools, created her mental schedule (complete with a deadline), and got to work. She was efficient, frugal, and tireless. She didn't waste time whining or complaining, and expected children to follow her example.

Lena found satisfaction in completing each task to the best of her ability. She was an efficiency expert who taught me the wisdom of cleaning the kitchen one more time immediately before going to bed. Unless a late-night eater foiled her plan, the kitchen would be clean in the morning, and breakfast preparation could begin without delay. She taught me that keeping rooms tidy conveys an appearance of order and cleanliness even if we didn't have time for what she considered deep cleaning.

Although Lena didn't use her early training as a nurse to pursue a career, she and those around her did benefit from her nursing skills. Through her knowledge of disease prevention and treatment combined with her ability to perform simple medical procedures, her children all grew to adulthood and she and Marvin lived long, healthy lives. She was pleased to assist her neighbors during childbirth, and felt especially close to children she'd helped enter the world.

In August 1960, Lena was putting bowls of food on the kitchen table one evening at suppertime when an approaching thunderstorm produced a lightening strike. The bulb over the table exploded. Lena was busy throwing away food spoiled by glass slivers, and wondering what she might serve for dinner, when a neighbor, Miss Flossie, came hurriedly into the kitchen to inform Lena her house was burning. According to Miss Flossie, a fireball had rolled along an electrical line and burst into flame at the upstairs corner of the house. Flames quickly engulfed the solid pine structure.

As usual, the town of Bonlee came to the family's aid. Men leaving their jobs nearby came to the rescue, removing much of the furniture and belongings from the ground floor, while volunteer firefighters held the flames at bay. Several volunteer fire departments responded and worked tirelessly for hours, yet they were unable to save the house. Everything on the second floor was lost, including family photographs and other irreplaceable items, but everyone was alive. Lena and her family lost the house they called home for forty years, but memories born in the rooms and on the porches were indestructible.

Their neighbor, Ralph, who had an empty rental house at the time, directed men to load their trucks and move Lena and Marvin into his house. By bedtime, Lena had set up housekeeping in their temporary home. Men had even moved the refrigerator without breaking an egg. As soon as the embers cooled, neighbors began helping clear the site, then donated materials and labor to help Marvin and Lena rebuild. Within a few months, they were living in a new brick home. Always trying to look on the bright side, Lena said she was pleased with the reduced effort of maintaining a smaller, one-story house. She thought it oddly coincidental her adult home was lost to fire precipitated by lightening, and her father's home met the same fate half a century earlier.

Lena seldom mentioned material possessions destroyed in the fire, although she was sad to have lost one-of-a-kind photographs that were precious to her. As usual, she looked to the needs of those around her then replaced an item she deemed important. Although my mother and I had moved away from my grandparents' home prior to the fire, I spent most weekends with Lena. Since I departed for church from her home, I kept my Sunday school Bible in Lena's bedroom, and it burned in the fire. The first Christmas following the fire, Lena purchased a Bible similar to the white leather-covered Bible my church had given me. On the dedication page inside the little Bible, she wrote that "Mama Lena" presented it to me on December 25, 1960. Holding that Bible half a century later, I realize how wise Lena was to chose a gift of lasting value.

In October 1961, Lena and Marvin's children hosted a fiftieth wedding anniversary party for their parents. Held at *The Elms*, a lovely tearoom in Siler City, they welcomed local friends as well as extended family members and friends from Pleasant Garden, where they had married half a century earlier. Although Lena's first marriage was tragically brief, her second one endured.

Two events within a two-week period in March 1963 left Lena and her family reeling. Marvin died suddenly of a heart attack on March 11. Lena's widowed daughter, Marie, died in a tragic automobile accident on March 23 leaving her teenaged son without parents. Again, Lena focused on the living. Lena, the emotionally stronger parent, was thankful Marvin escaped the pain of losing another daughter.

In 1974, the family's grief again came in double portions. On February 5, 1974, Lena and Marvin's eldest son, Louis, died suddenly of a heart attack. Two days later, his wife, Dorothy, collapsed on the funeral home steps and died two hours later. Three more of Lena's grandsons had lost their parents.

The family's next loss was Lena. She died on August 20, 1977, at the age of eighty-nine. During her later years, when asked about her health and vigor, she attributed her longevity to keeping her non-pregnant weight at 132 pounds, eating healthy food, and getting exercise from keeping house, walking to the grocery store, and working in her yard and garden. Her only long-term physical issues had been high blood pressure and arthritis in her knees. When she began to have little strokes that made her unsteady on her feet, several falls necessitated constant supervision. She spent her last months in a nursing home. During one nursing home visit, many of her grandchildren and great-grandchildren crowded around while she amused us by remembering and singing all the verses of *Froggie Went A'courting*, a folk song she had sung to us repeatedly decades earlier.

Lena outlived two husbands, four of her nine children, and her first grandchild. Since Lena's death, her remaining five children have passed away. Her son, John Owen, died suddenly of a heart attack in 1981 followed by Grace in 1982. Evelyn lost a lengthy fight with cancer in 1984. Lena's son by her first marriage, Robert, died in 1984. Carol, Lena's youngest daughter, died in March 2008. In February 2011, all ten of Lena's later grandchildren were living, and had added sixteen great-grandchildren, and four great, great-grandchildren to her descendants.

During Lena's life and following her death, substance abuse and physical abuse, similar to issues described in Grace's story, occurred among some of Lena's other children and their spouses. Lena, troubled by addictive behavior of her loved ones, attempted to reason with them. If they initially refused her counsel, she prayed for them, and remained available to offer adults and children a meal or a safe place to stay. Sadly, since her death, a few of Lena's grandchildren and great-grandchildren have struggled with addictions. True to His word, however, God has repeatedly worked miracles in the lives of those who were willing to turn to Him. Some who have recovered, lead ministries to help others move from addiction to freedom through faith in Jesus Christ.

Merely reading about Lena's life can make the most energetic person feel tired. How did she endure? From my first-hand observation, I believe her faith was her foundation. She also enriched her life through fulfilling activities she could afford on her meager budget. She never learned to drive a car. She was grateful for family members and friends who drove her to church, appointments, shopping, and meetings she wanted to attend. She attended church regularly. As long as her church had a choir, Lena sang with her neighbors and friends. What she lacked in quality, she made up for in volume. She prided herself in knowing many hymns by heart; enabling her to lift her eyes from the hymnal to look at the congregation while she sang. I remember being able to identify Lena's voice above the rest. In her youth, a visiting music leader taught her to project her voice and she never forgot that lesson. Lena sang to worship her Lord. When her voice reached His ear, it was a sweet sound.

Second only to camp meetings, Lena's enjoyed her Women's Missionary Union (WMU). Meetings were held in members' homes and, following the program, the hostess served an afternoon dessert tea. While enjoying a beverage and homemade treats, members deepened their friendships, and discussed local news. Because she spent her life preparing large quantities of basic food, Lena was not an accomplished cookie baker. She enjoyed collecting recipes from WMU members skilled at preparing dainty fare. Our favorite was Swedish Tea Cakes, a specialty of Lena's elegant, much-admired friend, Ellorie Pugh.

During her later years, when she was free from so many mouths to feed and had lighter household responsibilities, she enjoyed having more time to socialize with other women outside her church as well. She was a charter member of the Bonlee Women's Club. She attended Home Demonstration Club meetings where North Carolina County Agents taught homemaking skills. She implemented new ideas, tried new products, and enjoyed learning to modernize housekeeping and cooking methods. In addition to fellowship provided by those formal organizations, she seemed to relish the simple pleasure of chats with neighbor women who regularly dropped by after supper. She put her family responsibilities first; then, whenever possible, she availed herself of every social opportunity Bonlee life offered.

Lena enjoyed traveling and was eager to hop on a train to visit her children and new grandbabies. She visited children in Pennsylvania, South Carolina, New York, and Virginia. She visited her brother in the Virginia Beach area of Norfolk, Virginia. With more money and time, I believe Lena would have traveled the world. She was fearless and inquisitive. She enjoyed knowing people of all races, nationalities, religions, and viewpoints.

Although Marvin had an actual extra-marital affair, Lena admired a few special men from a distance. She had affairs of the heart we easily forgive. First, she adored Walter Cronkite and rarely missed his trusted CBS Evening News reports. As a young North Carolina woman who remembered hearing some guys named Wright were trying to fly out at Kitty Hawk, she said if Walter had not reported the story, she would not have believed men actually walked on the moon during her lifetime. Her other loves (although fictitious characters) were Perry Mason of the show by that name, and Matt Dillon of *Gunsmoke*. I say she loved the characters, because she admired the strong, successful men portrayed in the shows, and knew nothing of the private lives of Raymond Burr and James Arness.

Lena was an avid reader and kept a book next to her chair to read during any daytime moment she could spare. In the evenings, she took the book she was reading to her bedroom, where she would read a few more pages at the end of her day. Enjoying the mental challenge, she proudly completed the weekly newspaper crossword puzzle. If you were in Lena's household and wanted to stay out of trouble, you made certain she received the section of the paper where the crossword puzzle appeared. She didn't seem to need much in life, yet her few little pleasures were important to her.

I've mentioned Lena's corporate worship and involvement with her Women's Missionary Union as positive ways she endured her grueling, work-filled life. Lena's underlying individual strength came from her nightly habit of praying and reading her Bible. She only allowed herself the pleasure of reading the novel or biography she brought to bed from the living room after completing her daily Bible reading. She wanted to remain awake to finish her Bible passage then if she fell asleep, she only neglected her secular book. Simply by reading her Bible in my presence, Lena taught a valuable lesson.

When my mother and I first returned to North Carolina, my grandparents were still living in their large two-story home. With no central heating, wood-burning stoves and oil-burning heaters warmed the rooms in winter. To conserve resources, Lena, my mother and I shared a large bedroom on the second floor. My grandmother's small cot sat against one wall, and the double bed my mother and I shared was across the room. A vision of Lena lying on her cot remains clear in my mind's eye. Lena rested on her side with her head turned upward, held her Bible over her face, and read by the light of a small gooseneck lamp.

Curious, I often asked what she was reading. Never acting as though my interruption disturbed her, she lowered her Bible, looked at me, and recited a short version of the passage she was reading. Then, wanting to return to her reading without further interruption, she would say, "Now turn over and go to sleep."

Perhaps having had an empty nest for a few years before my arrival, Lena had hoped for time to herself. When my mother and I arrived, we disrupted her plans. Although I'm sure my presence did intrude on her private time, she welcomed me. Perhaps she was remembering a time when her own grandmother, Tabitha, had lived with Lena's family. Perhaps Tabitha shared favorite Bible passages with Lena. I'm grateful to have spent time with Lena when I was the only child in her household.

My cousin, Sally, who has one of Lena's Bibles, has explained there are no underlined passages, but Lena wrote "here" and "here" throughout her Bible. When Lena stopped reading each night, she wrote "here" to mark her stopping place and to indicate where she should begin reading the following evening. Some people are horrified that anyone dares to write in a book (especially a Bible), and I agree if the book belongs to a library and not to the individual. I'm pleased several of my ancestors penned evidence of their Bible study and prayer lives in Bibles, prayer books, and devotional books.

Let's briefly recall the major events of Lena's eighty-nine years. She lost her mother when she was nine and was keeping house by age twelve. When she was seventeen, her six-year-old little sister died. At eighteen, she was happily married and conceived her first son then lost her husband after only seven months of marriage. She gave birth to her first son when she was nineteen. She raised her son in her father's household for nearly five years. She married again when she was twenty-three and between 1911 and 1927, had eight more children. While her family was growing, she lost her mother-figure grandmother, Tabitha, her brother Banks died of TB, and she lost her father. She then endured the death of her second husband, four of her nine children, and her first grandchild.

Lena was highly successful to survive all that happened in her life and remain sane. She did, I must pause to admit, have two substance vices. The first I'll describe with a smile and the latter with a grimace.

First, she loved Dr. Pepper sodas. She did not often keep them in her home; rather, she enjoyed a soda when she visited her friend, Ralph Brooks, at his grocery store. When she entered Ralph's store, he seated her behind the counter then opened a cold little bottle of Dr. Pepper and handed it to Lena. Ralph then proceeded to fill her grocery order, exchanging pleasantries with Lena each time he returned to the counter with a grocery item. I believe Lena wanted to adopt Ralph. She loved that diligent man who called her Miss Lena, made her feel so welcome in his store, and supplied her Dr. Pepper fix. Ralph provided the temporary home for Lena and Marvin when their home burned and regularly hauled Lena's trash away. When she could get persimmons, she returned Ralph's love in the form of homemade persimmon pudding.

Lena's second vice was snuff. She also purchased her snuff at Ralph's store. Her snuff came in little beige paper-wrapped bales. Dipping snuff means, of course, the person doing the dipping will eventually need to spit. Ugh. The household chore I always tried to avoid was cleaning the *spittoons* used by Lena, and by my uncles, as receptacles for the spit from her snuff and from their chewing tobacco. Snuff was part of country life in North Carolina although I never understood the allure.

Thanks to Cousin Charles, we have an outsider's view of Lena. He provided a poem written by Ina Dunlap Andrews, Lena's best female friend and fellow Bonlee Woman's Club member for decades. When I think of those two women, I'd describe them as *no nonsense* women. They had a tireless, uncomplaining work ethic in common. Miss Ina taught Lena's children in school, and then was a substitute teacher in my own era. I knew Miss Ina best as a stalwart member of Bonlee Baptist Church, so I know her words to be sincere. Miss Ina wrote that Lena had never given up during difficult times, worked hard, yet was good-natured and jolly. "She asks no pity and wants no sad talk as she pushes the extra mile," Miss Ina wrote of Lena in her lengthy poem. She referred to Lena as her loyal and brave pal, then closed with the exclamation, "Oh, what a Gal!"

One family issue, however, looms large: Lena apparently failed to encourage the spiritual growth of her offspring. She took her children to Sunday school and church when they were young; however, she said she didn't talk to them about faith. She exposed children to spiritual truths, and then allowed them to find their own way, in their own time. Lena provided thousands of biscuits and miles of cornbread yet her children hungered for the Bread of Life.

The Biblical book of Proverbs tells us the mother who doesn't teach her child is contributing to the child's failure and even to the child's death. Lena's children needed spiritual discernment to choose appropriate spouses and friends, to shun bad lifestyle choices, and to attain emotional stability. Lena's example positively influenced me, and I'm grateful; but her children needed loving spiritual intervention. Outsiders or other family members have led her descendants to the Lord or they have remained lost.

I'm certain Lena prayed for all her children and grandchildren, and I'm a grateful beneficiary. If you haven't done what you feel you should or could do to help your children and grandchildren grow spiritually, prayer is an excellent starting place to become empowered. Don't stop at prayer. Go the distance.

The name Lena's father originally chose for her, Selena, means *of the moon*. Our moon generates no light of its own; it appears to have light by reflecting the light of the sun. I have a clear memory of Lena, reading and praying in the circle of light from her gooseneck lamp. Don't be remembered for being spiritually strong in a small circle of light. Be brilliant. Reflect the Light of the World: Jesus.

Marie

1893 - 1982

My father's mother, Marie, was a petite, gentle, well-mannered Southern lady who invoked the love of her family, and the admiration of friends and coworkers. As the only child of Marie's only child, I received her lavish, undivided grandmotherly love. A prolific correspondent, Marie closed her letters with the word *devotedly*, and she was true to her word. Marie was fully devoted to the welfare and happiness of her loved ones, even when her husband, son, and circumstances made her own life extremely difficult. God was the source of her strength.

Named for her father's mother, Marie Therese, Marie was the second child born to Daisy, of Maryland's Eastern Shore, and Vincent, a busy Methodist minister from Richmond, Virginia. Marie's life began on December 17, 1893, in the home of her maternal grandparents, Luther and Sarah, in Madison, Dorchester County, Maryland. Vincent was serving in Fox Hill, Virginia, but Daisy came home for her mother's help with toddler Vincent Jr. and infant Marie. Other siblings were Emma, born in August 1896; Grace, born in May 1901; and Walter, born in March 1906.

Her father's parents, Tony and Therese, lived in Richmond, Virginia. Marie developed a close relationship with her paternal grandmother, Therese, and lived with her the winter of 1907. During that visit, Marie copied a poem that hung on the wall over Therese's bed until her death.

Marie's father, Vincent, inherited the poem and hung it over his bed where it stayed until he died. Marie mentioned it to me and recorded it in her devotional book, so she apparently found it reassuring. The poem, titled *Good Night,* reads:

Sleep sweetly in this quiet room, O thou, who 'ere thou art
And let no mournful yesterdays disturb thy peaceful heart
Nor let tomorrow scare thy rest with thoughts of coming ills.
Thy maker is thy changeless friend. His love surrounds thee still.
Forget thyself and all the world. Put out each feverish light.
The stars are watching overhead. Sleep sweetly then, Good night.

Marie and her siblings visited Luther and Sarah in Madison each summer where one of Marie's playmates was Emerson, a boy eight months her junior with whom she had much in common. Both could trace their roots to early families of Maryland's Eastern Shore. Emerson's parents were natives of the same Maryland County. Emerson's father, a great political leader in their state, traveled to serve the public. Marie's family moved frequently to locations of her father's church assignments in Maryland and Virginia. Both families educated their children well.

Marie graduated from Richmond Normal School (high school) in 1912. On June 26, 1912, Marie met a young man named George while visiting an area called Atlantic City (now Ocean View) near Norfolk, Virginia. George was good looking, smart, and his father was a Methodist minister. Soon George proposed marriage and Marie accepted. Marie, who would soon go to teacher's college, agreed to wait three years while George, a Randolph-Macon College graduate, attended the University of Virginia law school.

Later that summer, Marie and Emerson met again at Madison. Emerson asked Marie to break her engagement with George, and wait while Emerson went to law school, then marry him. She said she'd already accepted George's proposal and planned to go ahead with her marriage. We'll leave Emerson for the moment and catch up with him later.

Marie wasn't idle during George's law school years. She enrolled at the Methodist-affiliated Blackstone Female Institute (later Blackstone College) where she earned her teaching credentials. At Blackstone, she endured peer teasing. The school had purchased bathroom fixtures from her Richmond grandfather's plumbing supply business with her family's surname stamped in the toilet bowls. Imagine the snide comments she endured.

While Marie was at Blackstone and George was in Charlottesville, their schools were ninety miles apart, and neither of them had money to travel. I imagine they exchanged letters frequently. When Marie graduated from Blackstone in 1914 and began teaching, she thought her brief career would end when she married. She later advised young women to prepare for an income-producing vocation even if they planned to marry rather than have a career.

George received his law degree in June 1915 and married Marie on November 24, 1915. Absent a wedding photograph, I have a photograph of Marie taken about the time of her marriage. Beautiful Marie, photographed smiling winsomely, wore her brown hair in a chin-length bob with soft curls framing her pretty features and blue eyes.

Initially, both families thought George and Marie a well-matched couple. Both families had similar geographic, religious and educational backgrounds. Their minister fathers worked diligently for little financial gain. Their experience as minister's children differed, however, because George's father had served established churches that paid better salaries and Marie's father served new or growing churches at lower salaries. George had one brother and Marie had four siblings.

On June 5, 1917, about a year and a half after George and Marie married, George described himself when he completed a World War I registration card. He was twenty-five and working as an attorney and land expert. George further stated he had not served in the military, and requested exemption from service because he was married, his wife was dependent upon him for support, and he was a railroad employee working under the Federal Valuation Act. He described himself as tall and slender with grey eyes and brown hair.

George appears to have been diligent while he was in college and an energetic career man who had more than one job. In November 1919, Portsmouth elected George (a Democrat) to represent them in the Virginia House of Delegates. Initially impressed at George's election, I learned he had run unopposed.

Before George went to Richmond to serve in the House of Delegates, the 1920 census captured a description of his Portsmouth household on January 10, 1920. George and Marie (seven months pregnant with my father at the time) were living in a boarding house at 500 High Street where George's older brother, Robert, was also boarding. The brothers, according to city directories for Portsmouth, were then partners in an automobile business. Marie lived about a mile from her parents' home near Wright Memorial Methodist Church, where her father was pastor at the time. Her in-laws, John and Mamie, lived about ten miles away in Norfolk, where John was pastor of Larchmont Methodist Church.

When I researched George's freshmen year as a 1920 Virginia delegate, I finally saw his adult face. Although Cousin Harry had provided a family portrait that included George as a young boy, I'd never seen my paternal grandfather as a man. In 2004, I visited the Library of Virginia in Richmond where photographs of members of the House of Delegates are stored. After I provided George's name and the year he served, the librarian located his photograph. George was a handsome man, so I could see why Marie was attracted to him. To me, George was a stranger yet so familiar.

Shortly after George sat for that photograph, Marie gave birth to a son on March 17, 1920. They named the boy John after George's father. The first (and ultimately only) grandchild for George's parents, John was the second grandson for Marie's parents. Marie began raising her baby boy in Portsmouth while George continued his busy career as city attorney in Portsmouth and served in Richmond. George's House of Delegates record indicates he served on several committees in 1920: General Laws, Public Property, Currency and Commerce, and Insurance and Banking. During his political career, George's list of memberships and activities included: lawyer, World War I veteran in the Red Cross service, member of the Methodist church, member of literary and debating societies (he won medals on debating teams at college and university), member of Sigma Phi Epsilon, and city attorney for Portsmouth. On paper, George was successful, hard-working, smart, good-looking, well educated, and a Christian. What is the downside of being married to George?

On July 4, 1921, George, Marie, and 16-month-old John went to visit George's parents in Norfolk for the day. That afternoon, George said he going out to purchase cigars and walked out the door. George never came back. He disappeared.

What did people think of George's disappearance? Marie never wanted to talk about it. Even fifty years after George left, her pain was evident. Cousin Bob in Baltimore found a newspaper article placed by the Elks Club that described George as a *brother* of that order. The article, published in Iowa on July 27, 1921, asked Elks to "find trace if possible" of their brother Elk who had disappeared from Portsmouth, Virginia. The article described George as a leading attorney, and then mentioned the Elks exalted ruler thought George demented, adding, "A lunacy warrant has been issued."

Did George leave voluntarily or was he a victim of foul play? Speculation has ranged from George's involvement in shady dealings, to George's possible abduction by persons involved in crime discovered by his House of Delegates committees, to George being ill. George's mother, Mamie, had Parkinson's disease so perhaps George inherited a genetic predisposition to that disease. Problem behaviors such as impulse control, psychotic symptoms, depression, and anxiety might accompany young-onset Parkinson's disease. We've criticized George for leaving his family yet he might have left against his will. He could have been mentally incompetent. Even though Marie never wanted to talk about George or her years of struggle, when her son began to exhibit violent behavior toward his wife, Marie said, "He is just like his father."

Abandoned by her husband of five years, Marie had a baby to raise, little money, and no job. Marie was soon able to teach school, because she moved into her parents' home where her mother provided daycare for John. Marie's parents were struggling financially, so Marie's salary helped defray her parents' household costs. Changes were occurring in the lives of Marie's siblings as well.

Marie's older brother, Vincent, had married a beautiful North Carolina girl in 1917 and they welcomed a son in 1918. Her younger sister, Emma had married in 1920 and on February 19, 1921 had given birth to a daughter she named Grace in honor of the next younger sister in the family. Six weeks after Emma gave birth, she died of double pneumonia. Marie had lost her first sibling.

Emma's husband, feeling ill equipped to raise an infant, left baby Grace in the care of his wife's parents, Daisy and Vincent. Five months later, Marie's husband disappeared and the household gained Marie and

John. In addition to her youngest children, Grace and Walter, Daisy had two infants in her care. The family members worked together and found ways to manage. The babies, John and Cousin Grace, felt loved and cared for in Daisy and Vincent's home.

The reaction (or lack of action) on the part of Marie's in-laws is puzzling. John was their only grandchild yet he spent very little time with his paternal grandparents. John's paternal grandfather appears to have ignored his namesake. John remembered his paternal grandfather as a warm, loving man and his paternal grandmother as a cold, formal woman. We'll revisit John and Mamie in Mamie's chapter; but with respect to Marie, their apparent lack of support is inexplicable. By 1930, Marie and John moved to an apartment in Richmond. Even though she was living on her own, the death of Marie's father in 1937 meant an important person in her support system was gone.

Marie's teaching credentials from Blackstone no longer sufficed. During summer breaks from teaching and in the evenings, she continued her education by completing two semesters at the University of Richmond (where she received an undergraduate degree) followed by further degrees from the University of Virginia at Charlottesville and Columbia University in New York City. Apparently, there were years when she had no one to care for John so she took him to the college she was attending. While Marie attended class, she made John (then twelve to fifteen) sit outside on the steps of the building. He was not to move or speak to strangers. In later years, Marie and John spoke of those summers as unpleasant times. Marie continued to do what was required to make a living and raise her son.

Marie's teaching career, necessary to support herself and her son, became a job she loved. Her teaching style reflected her gracious persona. Rather than speaking loudly to capture the attention of children who were disruptive, she spoke in a near whisper. Curiosity drove the children to silence so they could hear what she was saying. She controlled her classroom without giving up her ladylike demeanor.

Marie received gifts, cards, and notes from her students and their grateful mothers. She accumulated an impressive array of pretty handkerchiefs and aprons in addition to consumables such as cookies, candy, perfume, guest soap, and hand lotion. Marie, grateful by nature, raved over the smallest gift. Her love for others, her kind manner, and her patience were assets during her teaching career.

My love and respect for Marie didn't permit me to press her for details about raising my father alone when it was evident she was pained to remember those difficult years. My father was seldom willing to speak about his childhood or his relationship with his mother. One positive story from my father's youth illustrates his desire please his mother and follow her instructions with respect to good manners.

During a day of shopping in Richmond, Marie and John stopped for lunch in a department store tearoom. They ordered dessert (a rare treat), and then one of Marie's female friends approached their table. Since the visitor was female and his elder, John stood to greet her. When the woman refused an invitation to sit and remained standing, John obeyed his training and continued to stand. The waitress delivered dessert, and still the woman stood at their table, talking to Marie, while John's ice cream melted. Even when the woman told John to sit down and eat, he remained standing. When the woman finally left, John's ice cream had completely melted. Marie congratulated John on his fine manners and ordered fresh ice cream. In every life, there are good moments and stories with pleasant endings. Both Marie and John enjoyed re-telling the ice cream story.

At sixteen, John became embroiled in a matrimonial problem. A woman, believing John's family had money, encouraged her thirty-year-old daughter to entice John into marriage. Marie had the marriage annulled. John's ill-conceived marriage was the first of many unpleasant situations John created and Marie had to rectify. He was a shadow over every opportunity for her happiness, yet she remained a loyal mother.

Marie's next challenge involving John came when he decided to drop out during his senior year of high school to begin working for a radio station. A career in radio did fit his skill set; however, dropping out of high school was rebellious and unwise. Smart and well read, John attended college for a short time, then said the professors were stupid and didn't continue. Marie believed John's rebellion stemmed from the absence of his father.

In his late teens, John visited his mother's brother, Vincent Jr. and his wife, Gena, in Greensboro, North Carolina. John adored his aunt Gena and enjoyed Greensboro so much he wanted to live there. Even though

Aunt Gena passed away in October 1939 and Uncle Vincent moved back to Richmond, John was determined to return to Greensboro. On July 3, 1940, John left his mother's home in Richmond, went to Greensboro, and secured a position at a Greensboro radio station. On the July 3 page of her daily devotional book, Marie wrote, "John left this a.m. (1940). George left July 4, 1921. Both have left me."

Marie's devotional book is a great source of information about Marie's life and spiritual condition. The book, *God's Minute*, offers a prayer and sixty-second devotional reading for each day of the year. According to an inscription in the book, Marie received it as a gift from her friend, Cleve, in August 1939. According to Cousin Grace, Cleve was fellow teacher of Marie's, and the two women were great friends for decades. Based on the many notes Marie penned on the pages, the little book was frequently in her hands. Knowing she spent time in prayer helps me understand the source of her strength. Her personal notes offer clues about events of her life and her reactions. Marie also recorded first names of family members on the dates of their birth. I'm thankful she left her precious book to me, because it illuminates her faith during life's joyous and difficult times. Her notes and comments are included in the remainder of her chapter.

Even though her husband and son had left her, Marie never seemed lonely although some of her notes dispute her appearance. She had dozens of cousins in Richmond who kept in touch and visited as often as weekly. Sunday afternoon visits provided fellowship, and helped Marie stay abreast of what was happening in the lives of friends and family.

Marie recorded sad notes in her devotional book when she returned to Columbia University in New York, in July 1941 and July 1942 to continue her studies. The first entry reads, "July 31, 1941, at Columbia Univ., I work so hard and am so depressed." The next year, on the facing page, she wrote "July 30, 1942, Columbia Univ., What does it get me?"

Marie later told her daughter-in-law, Grace, that when John announced his intention to marry in 1942, she was horrified. The memory of John's first mistake with matrimony must have come to mind. Marie wondered if John's bride was another poor choice. Alternatively, she was concerned John's bride might be a lovely person whose life John would ruin. In October 1942, Marie stepped off the train in Greensboro, met Grace, and immediately knew John had found a gem. Marie later wished she'd warned Grace. The wedding took place and Marie prayed for her son and his new bride.

Even though Marie's relationship difficulties continued with her son, she felt blessed to have gained a wonderful daughter with whom she enjoyed mutual love and respect. Marie continued teaching school in Virginia. John and Grace began their married in North Carolina then moved to New York City in 1944. After I was born in 1947, Grace invited Marie to New York and offered to visit Marie in Virginia. Unpredictable John might happily meet his mother at the train station or at the last minute, refuse to see her. He might invite her to stay in our apartment or insist she stay in a hotel. Marie braved the wild fluctuations in her child's behavior to spend time with her grandchild and daughter-in-law.

Marie and I met for the first time in December 1947 when I was six months old. Too young to recall that first meeting, I have a photograph to prove we were together. Seated with me on her lap, Marie smiled prettily and I enjoyed a toothless, open-mouthed laugh. I'm so grateful for that perfect illustration of the joy Marie brought into my life.

A subsequent visit to New York, again over the Christmas Holidays, was awkward and unpleasant for Marie and Grace. The night before Marie's arrival, John insisted his mother was not to stay with us and shouted that he did not want to see her at all. Before he was finished, he had thrown our fully decorated Christmas tree to the floor, breaking many ornaments and lights. Grace stayed up all night to clean up the mess, rectify the tree, and book a room for Marie at a nearby inn.

Meeting Marie at the train the following day, Grace had to tell Marie that John refused to see her, and then we took her to the inn where she was to stay alone. When John went to work the following morning, Grace and I took Marie sightseeing and to lunch. Understandably, Marie shortened her stay and soon returned to Richmond. John's unkind treatment hurt and embarrassed Marie. After that fiasco, Grace decided visits with Marie should take place in Richmond. Marie was delighted. Photographs prove I visited Marie in 1950 but I was too young to recall our time together.

In the summer of 1953 when I was six years old, Grace took me to Richmond by train, and left me with Marie for a week. Marie gave me a lifetime of memories in one week by orchestrating every moment. Marie and I tied on pretty aprons and prepared our favorite treats. We ate our meals and held fancy tea parties at her tea table appointed with dainty china, silver, and embroidered linens.

Marie polished my table manners during our meal. I adored her so I wanted to mimic her every move. I learned to use an individual butter spreader and individual bread plate. She taught me to use a tiny silver shovel to scoop salt from a glass saltcellar. She made manners fun by teaching me little rhymes. My favorite was, "As little ships go out to sea, I dip my soup away from me." She treated me to chocolate ice cream made in an ice cube tray. She demonstrated polite dinner conversation, and taught me to have a *pleasant expression* on my face.

Wearing our hats and gloves, we went on a no-cost educational shopping spree at Miller & Rhodes department store. Marie gestured to beautiful examples of feminine attire while teaching me the new words *lingerie* and *negligee*. I was enthralled when we had lunch in the Miller & Rhodes tearoom, while watching models twirl among the tables wearing the latest fashions. When we received our check, Marie taught me to tip people as much as we could afford, because they were working hard and needed every penny we could share.

Marie entertained me by playing piano and singing. She read stories, fables, and poetry aloud. We walked to Franklin Elementary School, and she pointed to the windows of the classroom where she taught for three decades. She took me to visit her best friend, Ada, a fellow teacher. Marie made a six-year-old child feel protected and loved yet never babied. It was a happy week. Marie seemed to enjoy entertaining me as much as I loved being entertained.

In 1955, Marie called with news. Remember Marie's childhood friend, Emerson, who proposed to her when they were teenagers? After Marie declined his proposal, Emerson married and enjoyed a wonderful marriage that produced two daughters and grandchildren. Sadly, his wife passed away.

About a year after his wife's death, Emerson telephoned Marie and traveled from his home in Cambridge, Maryland, to take her to lunch in Richmond. Apparently, they had a few such dates to renew their childhood friendship, or so Marie thought. Then he asked her to marry him. Shocked, she said she took him out into the sunlight, and told him she wanted him to see how old and wrinkled she was (she was sixty-two). Undaunted, he still wanted to marry her. Her absent first husband declared legally dead, Marie was free to remarry. She was to retire from teaching, move to Maryland, and enjoy marriage with a man who was devoted to her.

When Emerson proposed, Marie was petite and slim with silver-streaked brown hair, blue eyes, and fair skin. She purchased the smallest fitted dresses and suits, then had them tailored to fit her tiny figure. In public, she wore dresses or suits and high heels. At home, she might wear a skirt and blouse with a cardigan and flats. She accessorized with jewelry, scarves, gloves, and hats. Marie's makeup products included floral-scented face powder, rouge, lipstick, and a touch of mascara. Her overall look was feminine, understated, and conservative.

Marie and Emerson married in an intimate wedding at 10:30 a.m. on Tuesday, November 15, 1955. Grace and I were so pleased Marie was marrying a kind, loving man. Marie deserved to be happy. I don't know what my father thought of his mother's remarriage.

Marie had a wonderful new life. She moved into Emerson's lovely home, enjoyed his daughters and grandchildren, and felt welcomed by Emerson's circle of friends. Her new home in Dorchester County, Maryland, wasn't a new geographic location to her. She was living only a few miles from Madison, where she was born and spent happy childhood summers, in a county where her ancestors had lived since the mid-1600s.

There were some changes in store, however. Emerson was Episcopalian. Though Marie had worshiped primarily as a Methodist, she had no issue with joining Emerson's church. When asked what she liked about being Episcopalian, Marie said, "I like to genuflect." What a great answer. She enjoyed a physical gesture of worship. Marie felt at home in an Episcopalian setting because she had worshiped with her Episcopalian paternal grandparents, Tony and Therese, when she visited them during her youth. She enjoyed social occasions related to the church as well. A note she penned in her little devotional book read: "Went to the rectory. Today the *Women of the Church* met. The house is so nice since being renovated."

Emerson was a saint. He never came home without bringing a little gift for Marie. He'd walk in with a pair of stockings, a flower, or a little piece of candy. He was a man of good works in his all aspects of his life but God surely rewarded him for being a wonderful, loving husband to Marie.

Marie's life became active as she accompanied her lawyer husband to social functions in their town and to conventions in Baltimore and other cities. Emerson had turned her lonely retirement into a life of loving companionship. When they weren't traveling or attending functions, she felt so blessed to have his older daughter nearby with her two children who called her Aunt Marie. She adored them as though they were her own grandchildren and shared Emerson's delight in their company.

My first visit to Marie in her Maryland home was in 1959. I liked Emerson before I really knew him because Marie was so relaxed and happy. Emerson further illustrated his love for Marie by making her grandchild feel welcome. He introduced me to his family, took me to see the local sights, and treated me to a boat trip on the Choptank River.

Marie's difficult son, John, soon reappeared to disrupt her happy life. After Grace left John, he moved several times to work for radio stations around the country. Although he had largely ignored his mother throughout his adult life, he began working for the radio station in Cambridge where Marie and Emerson lived. When John didn't need his mother, he crossed the street to avoid speaking to her. When John was in need, he took his troubles to Marie and, because Emerson graciously chose to help, to Marie's kind husband.

To Marie's great embarrassment, John frequently became drunk, and then came to their home in the middle of the night. Eventually, Emerson allowed John to live in their attic room. During John's stay, Marie penned numerous notes about John in the margins of her daily devotional book: "John in trouble, my poor boy. How my heart aches over him. John is in such a pitiful condition. God help us."

When John entered another doomed marriage, Emerson helped him obtain a divorce with stipulations that protected John and shielded his mother's funds. Emerson deserves sainthood. To the best of my knowledge, Marie's dear husband never complained. I'm sure Emerson's family didn't appreciate the problem-making son of their father's second wife, and perhaps even resented Marie because she brought him into their lives. To their great credit, Emerson's older daughter and her son were especially kind to Marie and later to me. Bless them.

Not only did Marie continue to support her son when he was in need, she helped support his child. After their divorce, John offered no alimony or child support to Grace. Afraid requesting support might result in John's demand for visitation rights, Grace didn't ask. Marie knew how difficult it was to raise a child alone. Marie used part of her teacher's retirement income to purchase school clothes for me and help Grace with other expenses. Again, Emerson's involvement is clear; had he not provided for Marie's physical needs, she wouldn't have had discretionary funds to help her grandchild.

Marie lost some of her joy when Emerson's health began to decline. Marie noted Emerson's illness in her daily devotional book in May 1960. Additionally, in the 1960s, I received a note from Marie's mother, Daisy, in which Daisy mentioned Emerson's illness and Baltimore hospitalization. He had returned home, and Daisy wrote that she was so pleased with his recovery. Reading Daisy's words, I wonder if she wished Marie had chosen Emerson as her first husband instead of George. I know Daisy loved Emerson and appreciated the joy he gave her daughter.

Daisy passed away in March 1965 having outlived Marie's father by more than twenty-six years. Marie, busy trying to help keep her husband healthy, had little time to grieve. On October 24, 1969, Marie made a note in her devotional book that reads, "E in bed, Dr. came twice, blood pressure and heart again." In 1971, Marie's older brother, Vincent, died. Eight months later, Emerson died in January 1972. Marie had lost her husband and best friend. She was heartbroken and although she tried to be brave, the joy Marie radiated during her marriage to Emerson was gone. Marie moved to a nearby apartment so Emerson's daughters could sell their family home. It was the end of an era.

Visiting Marie after Emerson's death refreshed earlier memories. Her Cambridge apartment, filled with her Richmond furniture, closely resembled her 1950s Richmond apartment. When my husband, my mother, and I visited Marie in 1975, she was still trim, petite, and gracious although her health had begun to fail. By that time, John was working consistently and married a fourth and final time to a kind woman.

Marie began experiencing little strokes, falling, and failing mentally. Her faithful younger sister, Grace, moved Marie to a good nursing home close to her own home in Norfolk in 1977. After she entered the nursing home, Marie's mind gradually slipped further and further away. Over the years, she had suffered from shingles and a few mild illnesses, such as colds or flu, yet remained generally healthy. The last time I visited, I barely recognized her. The always perfectly dressed, poised woman I'd known was wearing a sweat suit and sitting slumped in a wheelchair. When I told Marie who I was, she sat quietly for a few moments, then looked up at me and said, "You don't look like anybody." Marie's body was alive; her mind was gone. She remained in the nursing home until her death on Christmas Day 1982 at the age of eighty-nine years, eight days. Her Virginia siblings gathered to bury her near their parents, grandparents, and great-grandparents in Richmond's Hollywood Cemetery.

Marie possessed the strength of character to remain cheerful and kind, always trying to make others happy and meet their needs. As nieces, nephews, great nieces, great nephews, in-laws, and cousins joined her family, Marie loved them all. She was a faithful correspondent who kept in touch with family and friends. She appreciated phone calls and made people feel welcome in her home, however humble.

What was the source of Marie's strength? On a little slip of paper inserted next to the July 4 page in her devotional book where she recorded George's abandonment, Marie had written a scripture from Isaiah: "Thou wilt keep him in perfect peace, whose mind is stayed on Thee." Marie was wise to turn to God during the difficult days of her life.

Marie's notes in her devotional book illustrate words she thought important enough to keep handy for continued reflection. Close to the front, she listed the components of prayer: adoration, confession, thanksgiving, supplication, and commitment. She then added, "Prayer (is) the most powerful form of energy that one can generate. Its results can be measured in terms of increasing physical buoyancy, greater intellectual vigor, moral stamina, and a deeper understanding of the realities underlying human relationships. If you make a habit of prayer, your life will be noticeably and profoundly altered. Prayer stamps with its indelible mark our actions and demeanor. Tranquility of bearing and facial and bodily repose are observed in those whose inner lives are thus enriched."

"Thank God more often than you breathe," was a quote she wrote next. She polished off her prayer-focused page with these words: "Don't be ashamed to acknowledge that you need divine help with human problems."

On the next page, she quoted a favorite scripture: "For we know all things work together for good," from Romans 8:28.

Under the topic, "For Bereaved Friends," Marie wrote what she might say of her life as she quoted scripture from 2 Timothy 4:7, "I have fought the good fight, I have finished the course, and I have kept the faith." Although she didn't quote it, I hope Marie read verse 8 of that Timothy passage about the promise of victory for all who wait with love for the Lord.

After Marie's death, I read her personal notes and learned the depth and duration of her struggles. I knew some of the facts of her life but didn't realize the depth of pain she suffered. Marie shielded her loved ones. She refrained from making negative comments about anyone or anything. I now realize how much she relied on God when people disappointed or hurt her. What a valuable lesson. We set an example of God's grace when we remain positive in our words and actions, or practice something as seemingly simple as a pleasant facial expression.

Marie faithfully read scripture, spent time in personal devotions, and valued prayer. Husbands leave and sons disappoint, yet God is always there to listen and give us hope. It takes a strong person to ask for help, and I'm thankful Marie left evidence her help came from God.

Marie's love and influence are priceless gifts to me. She stepped past a disappearing husband and a difficult son to offer her love to me. If your child is a disappointment, follow Marie's example and muster the strength to have a positive impact on your grandchildren. It isn't necessary to live near your grandchildren to be influential in their lives. Spend as much time with them as possible even if distance limits most communication to phone calls or e-mail. Thoughtfully make every moment count. Be a strong grandmother who blesses future generations with wisdom and faith. Little eyes are watching you.

PART III

Great-grandmothers

Annie Belle

1858 – 1897

Annie Belle was born on May 8, 1858, in Guilford County, North Carolina. She joined Julius, Emma, and Franklin, as the fourth child of her parents, Tabitha and Sanders. Because Annie Belle was the only name I'd heard this great-grandmother called, and Annie Belle appears on her grave marker, I assumed it was her full name. Expecting to find a toddler named Annie Belle in the 1860 census, I was surprised to find two-year-old *Isabel A.*; she appeared again by that name at age twelve in 1870. Annie Belle had begun life as Isabel Ann. Her name honored her father's great-grandmother, Isabel (meaning *Mother Nature's favorite daughter)*, and his mother, Ann (*grace*).

The first woman named Isabel in Annie Belle's family must have been quite a gal. By the time Annie Belle was born, there were numerous females in her extended family named Isabel and called *Ibby*. So many women and girls in the family were called *Ibby* that one man's will included the phrase, "all my granddaughters named *Ibby*." Clearly, it was wise to give Annie Belle her distinctive call name.

If Sanders described his childhood to Annie Belle, she learned he was born into a large family. By the time he was nine, his parents were both dead and he was living with an older brother. Although orphaned, Sanders was financially secure; he inherited his father's plantation. Before the American Civil War, Sanders had real estate valued at $2,000 and a personal estate worth $2,800. For his era, his finances indicate Annie Belle lived a comfortable middle-class farm life. In 1861, her parents completed their family with Robert, born a few weeks after Annie Belle's third birthday.

With five children and a wife to care for, Annie Belle's father tried to avoid fighting in the war, then finally enlisted (or perhaps was drafted by force) in 1864 about two months before Annie Belle turned six. When Sanders went to war, Annie Belle's mother ran the farm and kept their family fed and clothed. A detailed account of George's war experience appears in the chapter about Annie Belle's mother, Tabitha. In short, Sanders was wounded, hospitalized, captured, held in a horrible POW camp, and then released in June 1865. He survived, but came home wounded and ill. He died in August 1865 when Annie Belle was seven.

Women in Annie Belle's family dealt with war on the home front. One story she probably knew well involved an older Isabel's creative solution to a potentially dangerous problem. The Ibby of this instance had raised a large family. Several of her sons moved west to Indiana before the war began and joined the Union Army when war broke out. Other sons remained in North Carolina and joined the Confederacy. One day during the war, some of her Union sons arrived on her doorstep ragged, crawling with lice, and hungry. Of course, she took them in, clothed, and fed them. Before her Union sons could leave, she saw her Confederate sons approaching, and she knew they had orders to shoot Yankees on sight. Ibby hid her Indiana boys in the barn, then fed and clothed her Confederate sons who went on their way never knowing their brothers were in the barn. Ibby's allegiance to her children was unwavering despite their wartime choices.

Annie Belle's spiritual training began at home and continued at Bethlehem Methodist Church. Bethlehem, originally formed as a Baptist church, joined the Methodist Conference in 1825. Annie Belle's first childhood church, a log building with a balcony for slaves, was replaced by a new building constructed in 1868 when Annie Belle was ten years old.

Facts about Annie Belle's elementary schooling have remained illusive. The 1870 census record, for example, indicates her older brother, Franklin (then fifteen), and younger brother Robert (then nine), were both attending school while Isabel A. (Annie Belle) is *at home* at the age of twelve. Census records do indicate, however, that Annie Belle could read and write. Good schools were available to children in her native county well before Annie

Belle's time. The Methodist church built an "academy" in 1869 where Annie Belle attained the equivalent of a high school education.

The 1870 census lists Annie Belle, her widowed mother, and her brothers in the household of Alexander and Rebecca, who were the parents of Annie Belle's mother. Since George's parents had died when he was a boy, it was fortunate Tabitha's parents were living and could help raise the family. Annie Belle's sister, Emma, died between 1860 and 1870.

When the census taker visited their area in 1880, Annie Belle was twenty-two and still living with her widowed mother and three single brothers. Annie Belle did not have a formal occupation; however, I'm sure she was doing her share of the household work.

Annie Belle's grandfather, Alexander, died in 1881. In the absence of her own father, her grandfather had been the father figure in Annie Belle's life. Her grandmother, Rebecca, reportedly died shortly after Alexander passed away. So prior to Annie Belle's marriage, her supportive grandparents were gone.

On December 31, 1885, Annie Belle married Robert "Bob", a dentist neighbor. She was twenty-seven when they married and her husband was thirty-two. Annie Belle had chosen well. The township where both families lived bore Bob's surname because his father, Frederick, had been one of the wealthiest businessmen in their county. Frederick's pre-war $40,000 personal estate deflated considerably before Annie Belle's marriage to Bob, but the family's diligence continued. You'll learn more about Bob's family in the chapter about his mother, Mariah. Bob came from a Christian family who, by the time he married Annie Belle, also worshiped as Methodists.

Annie Belle and Bob worshiped at Pleasant Garden Methodist Church. It was the closest Methodist church to their home and Bob was a member there when they married. Names of Bob's extended family appear in early Pleasant Garden Methodist Church records as active members who contributed such assets as land for the first parsonage built in 1877. The pastor who performed their marriage was Rev. Caswell King who was not, at the time, pastoring at Pleasant Garden. Rev. King was Bob's brother-in-law.

Annie Belle's first full day of married life was the first day of 1886, an eventful year. Annie Belle became pregnant during the first few weeks of her marriage. She gave birth to a daughter on September 25, 1886, and the child died the same day. Five days later, Annie Belle's older brother, Cicero, died.

A happy event, the marriage of Annie Belle's younger brother, Robert, occurred on December 23, 1886. Annie Belle gained a sister-in-law, Donna. In October 1887, Donna and Robert had their first child and Annie Belle's first nephew, George. By the time of George's birth, Annie Belle was pregnant again. My grandmother, Lena, made her debut on Christmas Day 1887. I wonder if Annie Belle felt joy or fear. Lena, as you know from reading her chapter, did thrive. Heartache followed joy again when Annie Belle's brother, Franklin, died sixteen days later on January 10, 1888.

Three years and three days after Lena's birth, Annie Belle gave birth to a son, Banks, on December 28, 1890. Another son, Owen Frederick "Ted", joined the little household on October 10, 1892. On March 17, 1896, Annie Belle gave birth to another daughter, Eula Belle, her last child. Annie Belle's husband earned a living as a dentist and undertaker. They lived on a small farm where Annie Belle tended her garden, milked her cow, raised chickens in her yard, and sewed clothing for her family.

In a family photograph provided by Cousin Julia, Annie Belle is pictured with her husband, Bob, and three of their children; Lena, Banks, and Ted. The absence of Eula Belle, born in 1896, helped me date the photograph at about 1895. Annie Belle's facial features and dark hair made me wonder if she resembled her Scot-Irish ancestors. She appears slim, yet her face isn't as thin as Bob's, whose cheeks appear quite sunken. Perhaps he had already begun to worry about his wife's health.

In the photo, Annie Belle's dress is dark and simple with a small pin at the high neckline. Annie Belle's hands, slender and quiet in her lap, were probably seldom still with a family to feed and clothe and a house to tend. Her eyes appear to be dark blue or brown. Bob's eyes appear light like the pale blue eyes of two of his children, my grandmother, Lena, and my great uncle, Ted.

On October 9, 1897, Annie Belle died. How long she had been ill and how much she had suffered, my grandmother didn't recall. The first hospital had opened in nearby Greensboro in 1891; however, I don't know whether Annie Belle went to the hospital or remained at home. She died, as her father had, of tuberculosis.

Annie Belle's mother, Tabitha, moved in to help raise the children. The youngest, Eula Belle, hadn't reached her second birthday when Annie Belle died. Tragically, two more of Annie Belle's children died of tuberculosis. Eula Belle passed away at the age of six. Banks, who had become a semi-professional baseball player, died at the age of twenty-eight having never married.

My grandmother, who was only nine when her mother died, said she knew little about her mother. I suppose Annie Belle's premature death was tragic and caused my grandmother to shift her love to her grandmother, Tabitha, who came to fill the mothering role for the family and stayed three years.

When I've attempted to imagine Annie Belle's personality, I've thought she might have been spunky. My mind's ear can almost hear good-natured banter in my grandmother's account of Annie Belle's discussion with her husband about the naming of Lena. You read about that discussion in Lena's chapter and know the outcome.

Annie Belle and Bob had known each other for many years before they married; indicating a marriage based more on enduring friendship than romantic infatuation. Bob had been away to medical school in Atlanta and had opportunities to choose his wife from a wide spectrum of women. Perhaps Annie Belle's perceived qualities as a wife and mother brought Bob home to their small North Carolina community to marry her.

In an effort to understand Annie Belle's symptoms when she became ill, I searched for details about tuberculosis. It is likely George exposed Annie Belle and her siblings to TB when he returned from the war. Alternatively, Annie Belle's husband was a dentist who also performed tasks as an undertaker. Frequently exposed to infected individuals (living and dead), Bob could have passed infections to his family. After becoming consumptive herself, Annie Belle might have unwittingly passed TB on to her children.

Rural households in the late 1800s were likely small and not well ventilated. Airborne bacteria from an infected person's cough posed a threat to others in the home. Annie Belle's mother, younger brother, husband, and two of her children survived without developing TB. Perhaps they were carriers who resisted development of actual symptoms. Perhaps they escaped illness and death by the grace of God.

Imagine what Annie Belle endured. Within two to eight weeks of being infected, a person's immune system begins to attack TB bacteria. Some people are able to fight off the infection and kill the bacteria. Others experience no symptoms, although the bacteria can remain active in the individual's system for decades, enabling them to pass the disease to others without feeling ill themselves. TB becomes deadly when the individual's immune system is unable to fight off the infection. Exposed to TB either by her husband or by her father, the disease may have remained dormant until physical stress from childbirth and the demands of her busy life allowed the disease to manifest.

If Annie Belle did suffer, perhaps she did her best to hide the illness from her children, since she knew from her own experience the sadness of seeing a parent slip away. To minimize her own suffering and for the sake of her husband and children, I hope she died quickly once it was clear her illness was terminal.

Dying at thirty-nine, Annie Belle had precious little time to spend with her daughter, Lena, who is my connection to her. Did Annie Belle encourage Lena's daily Bible-reading habit? Perhaps Annie Belle read her Bible in Lena's presence, and read aloud to Lena before she could read herself. If she taught Lena to read, she might have used the Bible for part of those reading lessons.

Bob did not remarry after Annie Belle died, even though he survived her by twenty-three years. Perhaps Tabitha's willingness to help raise the children alleviated the pressure to remarry to gain a helpmate. Maybe he never got over the grief of losing his wife, or no other woman he met could take Annie Belle's place in his heart and home.

I've been unable to locate a Bible, letters, or hand-written notes attributed to Annie Belle. Two fires appear to have destroyed anything Annie Belle might have owned or written. After Annie Belle's death, lightening struck and fire consumed her husband's home. Precious family photographs and mementos that survived the first fire in the possession of Bob's daughter, Lena, were lost when Lena's home burned from a lightening strike forty years later. Lena mentioned losing items significant to her family without describing them in detail.

After investigating Annie Belle's parents and ancestors, from whom she received DNA and nurture to create the woman she became, we can observe her children and their actions for clues about her worthiness. Annie Belle's two surviving children ensured that at least part of her name lived on in the lives of her descendants. Lena named her first daughter Mildred Belle. Ted named his first daughter Mary Belle. Annie Belle's children barely knew her yet wanted to honor her.

Annie Belle's life brings a particular truth into sharp focus for us: the length of our life is uncertain. We should be prepared at all times for the next life because it could begin far sooner than we might have planned. The lesson we might learn from Annie Belle is to *be ready*. Each woman's life, however brief, can have a lasting, growing impact. Her two surviving children had produced over fifty descendants by 2011. Annie Belle's DNA has spread across five generations.

To summarize Annie Belle's spiritual life, we know her ancestors were Presbyterian and Methodist Christians. Annie Belle attended Methodist churches, a Methodist minister performed her marriage ceremony, and her grave is in a Methodist church cemetery. Many of her descendants are spiritually strong, Bible-believing Christians. Aside from those historical facts, I don't know where her heart was spiritually.

Standing by Annie Belle's grave in that pretty North Carolina cemetery, I've looked forward to the meeting we'll have in heaven. To paraphrase from 3 John 13 in the New Testament, I have much to tell her when we can talk face-to-face. For now, I trust Annie Belle accepted the gift of salvation and lives God's presence.

Julia

1849 – 1935

Let's catch a glimpse of Julia before we explore the events of her life. She died before I was born, so I'm fortunate my cousins saved photographs and penned descriptions of her. Cousins Frances and Katherine developed a terrific family book that includes Julia and her ancestors. Much of what you'll read in this chapter is attributable to their diligent effort and open sharing. You'll read more about that family book in my chapter about Julia's mother, Hannah. Frances and Katherine spent hundreds of research hours collecting family facts and photographs then organized them into one volume. Because of their efforts, I had no trouble finding photographs of Julia. Images of her appear in the pages of their book at two stages of her life spaced about a half-century apart.

A photograph taken of Julia in about 1869 captures a pretty, young woman of about twenty wearing a well-accessorized outfit. Seated with one arm resting on the arm of a sofa and one dainty hand up-turned in her lap, Julia's expression looks serious yet slightly coy. Her mouth and brows form straight lines as though she had to concentrate to sit still.

Although the photograph I received is quite small, high-resolution scanning and enlargement revealed clothing details. Her dress has the expected nipped-in waistline below a bodice embellished with ruffles and layered bows. A beautiful lace collar frames her face, and her long sleeves end in ruffles over lace cuffs. Topped with a jaunty little hat trimmed in flowers, her dark hair is swept off her face and hangs in sausage curls. In addition to all the heavy fabric, lace, and bows, she is wearing a ribbon neckband with a star-shaped pendant. Looking at Julia's youthful picture, the words *all dolled up* come to mind.

A woman of spiritual substance who was more than her looks and clothing, she was the mother of my grandfather, Marvin. You read briefly of Julia in the story of Marvin's wife, Lena (Julia's daughter-in-law). Now we'll capture Julia's life story and explore her father's ancestry.

Julia Jeanette was born on February 25, 1849 in Randolph County, North Carolina. The seventh child born to her parents, Hannah and Sampson, Julia joined three older sisters and three older brothers. Her mother later gave birth to two more sons to complete their family of nine. At the time of Julia's birth, her only living grandparent was her maternal grandmother, Ann.

Julia's middle name, Jeanette, doesn't appear to have familial origins. Julia's father had a niece named Julia. Sampson's younger sister, Susan, had named a daughter Julia in 1838. Sampson signed his sister's marriage bond, so perhaps Sampson and Susan were loving siblings. Susan passed away in 1844 so Sampson and Hannah may have been honoring their motherless niece. Julia means *youthful*. The meaning of Julia's middle name, Jeanette, seems prophetic upon learning the name of her eventual spouse. We'll revisit that meaning later.

To understand Julia, we'll step back in time to learn about the people who came before her. Her mother's bloodlines mingle with some well-known colonial American families, as you'll read in Hannah's chapter. The ancestral lines of Julia's father, Sampson, provide an earlier historical connection. Sampson's people came to North Carolina from Scotland where, if early genealogists were correct, they descended from Robert the Bruce, King of Scotland. Julia's twelfth great-grandfather reportedly married Margaret, daughter of Robert the Bruce and an unknown woman.

Why was Margaret born out of wedlock? Was Margaret's mother unable to refuse the king? However improper Robert the Bruce's initial act of infidelity, evidence indicates he took responsibility for his offspring. At least in her adulthood, Robert played a role in arranging a good life for his daughter. When she married a man her father held in high esteem, Robert the Bruce gave her husband land. Such a land gift meant putting Margaret's husband in charge of part of the country. Reportedly, the son-in-law (also named Robert, so we'll call him Rob), had been a soldier and close ally of the king. The land gift was likely also a reward for the son-in-law's faithful service. Perhaps the King was securing a place for his daughter in society by helping her marry well and, through the land gift, Margaret and Rob had the financial means for a good life for themselves and their heirs.

When researching our ancestral women, we often run into what genealogists refer to as *brick walls*. In short, that term means we have reached a person whose ancestry remains a mystery. While persistence usually helps us break through that wall to learn about the people on the other side, the challenge of identifying Margaret's mother and maternal ancestors is daunting.

At the time of Margaret's birth and childhood, people knew the identity of her mother. Aside from his legitimate heirs, Robert the Bruce reportedly fathered several children by *unknown* women whose names have remained a mystery. If historians had identified Margaret's mother, her name would appear in historic accounts of the life of Robert the Bruce.

How were Margaret and her mother treated in the early years of Margaret's life? Was Margaret taken away from her mother to live among the king's other offspring? I'll probably never know the name of Margaret's mother or her story. God knows her. I'm grateful a woman I can only call *Margaret's mother* has a place in the line of women who make my life possible. For now, I'll have to accept a blank pedigree where Margaret's ancestors should be.

We'll skip several generations and move our story from Scotland to America. Julia's paternal ancestors arrived in Virginia in the 1700s. By 1761, they were in Orange County, North Carolina. The first man of her father's line to arrive was a physician and minister before leaving Scotland, although I've found no evidence of him serving in either capacity after his arrival in North Carolina. By the time Julia's father was born, his people were farming the rich North Carolina soil.

Sampson's ancestors, determined to work hard and prosper, soon amassed twenty-five miles of North Carolina land. By the time Sampson's father died, the family had fallen on lean financial times. Sampson's siblings raised and helped educate him, but he received no land. As a young man, Sampson became overseer of another man's plantation. More about Sampson's life story appears in the chapter about his wife, Hannah.

Schooled at home, lessons for Julia and her siblings included good penmanship and music. Teachers were their mother, Hannah, and Virginia, Julia's eldest sister. Virginia's profession, "tutoress," appears in the 1860 census; she was then twenty-four and Julia was eleven. Perhaps Julia formed her habit of daily Bible study during her early reading lessons. In their household, everyone (including servants) gathered for prayer. When Julia and her siblings voiced education, Bible study, and prayer memories, they referred to their mother as the person who taught and maintained the vigil of those habits. Already, we see Julia receiving foundational spiritual training, thanks to Hannah.

In 1858, when Julia was nine, her family moved from her mother's family plantation in Randolph County, North Carolina to a farm in a more northern part of their county. Called Vickery Place, their farm was located between Centre Friends Meeting and Level Cross. The census taken in August 1860 notes the family living at Vickery Place, but Julia's maternal grandmother, Ann, was notably absent. Ann, widowed in 1835, lived with Julia's family until (after a five-day battle with pneumonia) she passed away in March 1860 at the age of eighty-eight. Ann had been a member of the Methodist Episcopal Church for forty years, according to her obituary. The chapter about Julia's mother, Hannah, will provide more insights about Hannah's parents, siblings, and their faith. From Julia's perspective, spiritually strong women raised her and surrounded her.

Julia's Quaker father, always working to advance his land holdings and the quality of his family's lifestyle, wanted to move his family closer to a Quaker meeting. By the early 1860s, he purchased land near Centre Friends Meeting in Guilford County. The new property included a house and barn.

Prior to the family's move to their new location, Yankee soldiers burned their barn. A family, observing the fire from their neighboring farm, dispatched a boy named John to inform Julia's family their barn had burned. Having walked about four miles to reach Julia's home at Vickery Place, John later said his reward for the effort was meeting Julia, then a barefoot girl with curly hair.

Julia was a teenager when three of her brothers left home to fight in the American Civil War. In North Carolina newspapers of the early 1860s, Julia might have read words such as *slavery is a sin* quoting abolitionists who agreed with her parents, next to an advertisement of *Negroes for sale.* Her father's name appeared in an 1861 newspaper upon his appointment to a committee charged with managing taxes earmarked for war costs. He helped make decisions regarding expenditures for clothing, arms, equipment, and training costs for North Carolina volunteer troops. He worked with his fellow committee members to advise the court with respect to providing loans for indigent families of men who were fighting or who had lost their lives. The topic of war and its associated hardships must have dominated their family conversations during those years.

Where were Julia's spiritual allegiances by the time she was nineteen years old? The record of her marriage to John indicates she might have learned something at knee of Ann, her maternal grandmother. A Methodist minister married Julia and John in her parents' home. To complicate the puzzle further, Julia and John appear to have raised their children in the Religious Society of Friends. Julia and John as well as their parents are buried in a Quaker cemetery. Many families I've researched in the community where John and Julia lived during their married life mirror their involvement in the nearby Quaker meeting as well as attending special events hosted by Methodist churches.

Well matched, Julia and John were both twenty-nine when they married in April 1878. They were both from families of nine children fathered by strong, opinionated men. Both had brothers who had fought in the American Civil War. Their mothers provided similar secular and religious education. You'll learn about the lives of John's parents prior to his marriage in the chapter about his mother, Miriam.

Let's take a second look at Julia's middle name, Jeanette. Usually spelled Jeannette, the name is a diminutive of the name Jeanne, the French form of the name Joan (the feminine form of the name John). The meaning of the name John is: God is good. I doubt her parents knew the root meaning of the name. They certainly didn't realize they were giving her the feminine version of her future husband's name. God works in mysterious ways.

When Julia married, she did not have a living mother-in-law, although she had known Miriam during a decade as her near neighbor. Julia's father-in-law, Andrew, lived nearby during the first twenty-five years of Julia's married life until his death in 1903. Andrew was an influential father and grandfather. Census records indicate Andrew's household often included adult children and grandchildren. Unlike Julia's siblings, many of whom moved far away during their adult lives, John's surviving siblings remained close geographically, and raised their families in the same county.

John and Julia set up housekeeping in a two-story wooden farmhouse. Julia's mother, Hannah, lived nearby during the early years of Julia's marriage and the births of her four children. Julia's first child and only son, Marvin "Pat" Baxter (my grandfather), was born on December 24, 1878. After Marvin, Julia gave birth to three daughters. The first was Miriam "Mamie" Moss, born in May 1881, and named for John's mother, Miriam. Her middle name, Moss, was an ancestral surname from Julia's family. Lillian Lucille Virginia was born in January 1883 and called Lillian. Except for the name of Julia's eldest sister, Virginia, who had passed away in 1881, Lillian's name has no family significance. In 1885, Julia gave birth to Hannah Mozelle (called Mozelle), her last child. Mozelle's first name, Hannah, honored Julia's mother and other ancestral women named Hannah.

John supplemented the family income by working as a court clerk and as a revenue agent, a job that might have given Julia cause to worry. On December 11, 1890, a newspaper article reported John had been appointed assistant revenue agent in Guilford County because another man had resigned. For Julia's sake, I hope John had a desk job. If John's position involved fieldwork, it was a dangerous job. Eugene, John's brother and a revenue agent, died on June 24, 1890 of injuries sustained during a distillery raid. John lived until 1932 so if he didn't have a desk job, at least he survived revenue raids.

Returning to the subject of church, following their wedding performed by a Methodist minister, family lore states John and Julia worshiped at Centre Friends Meeting. You'll learn in Hannah's story that one of Hannah's brothers became a Methodist minister as did Hannah's son, Edwin (Julia's younger brother). Of Julia and John's four children, two continued in the Religious Society of Friends (although Lillian worshiped with her Catholic husband during her marriage) and two became Methodist.

A few of Julia's descendants still follow her Quaker faith; others worship as Methodists and in other protestant denominations. If Julia could speak to us today, I'm certain she'd be happy with any denomination based on literal Bible interpretation and faith in Jesus Christ. She'd probably be delighted to learn a great-granddaughter served as an elder in her church, and even more pleased to learn that same woman and her husband have now moved to a denomination more accurately interpreting and following Biblical principles.

What I've learned about Julia's personal faith is from family lore and first-hand inspection of her Bible. As I took her Bible from a cousin's hand, I was thrilled to see she had underlined verses. I first thought her markings indicated Julia's favorite scripture verses. In a few moments, I realized she had underlined single verses and whole chapters in more than thirty passages. Here are a few highlights: Julia marked the story of the creation of man in Genesis, she marked twenty passages in Psalms, and she marked Proverbs 30:5 that mentions the flawlessness of God's word. The last passage she marked was Revelation 3 where she underlined the entire chapter. She must have been striving to comply with a statement she penned in her Bible: "Read a portion of God's word every day and abide by it." What a great testimony to her diligent study of the Word.

Julia's adult prayer life mirrored the prayer life of her mother, Hannah. The two women's prayers were evident, frequent, and fervent. During her married life, Julia made morning visits to a little closet under the stairs in their farmhouse, and prayed loudly enough for people outside the closet to hear her petitions.

People who knew Julia said she had a pretty face and twinkling eyes even in her later years when her skin had aged into hundreds of tiny wrinkles. The brown curls of her youth turned into pretty, white hair. Children who visited her household were amused when she lost her pocketbook and asked them to aid in the search. Relatives and friends of all ages said they loved her.

Why did Julia marry her neighbor, John? Described as a distinguished man of medium height, I didn't see John's face as a boy or young man until 2010 when I received a box of unidentified photographs from Cousin Martha. I knew John attended a Quaker school in Indiana in 1870. As I began to identify photographs of people I knew to be in Indiana during that timeframe, I found a photograph of John when he was about twenty-one. Handsome in his white shirt, bow tie, and dapper jacket with contrasting piping around the lapels, John has a pleasant expression. His eyes droop slightly at the outside corners, are rimmed in dark circles (a family trait), and he sports a wispy mustache.

Beneath John's outward appearance, he had a photographic memory and a witty, teasing personality. According to a letter written about 1880 by Julia's oldest sister, Virginia, Julia had many funny things to say. She and John had well-matched, fun-loving personalities.

Nearly half a century after Julia sat for her youthful photograph (described in the opening of her story) and John had his Indiana photograph taken, they were photographed together in 1918 when they were sixty-nine. Seated in wooden chairs placed side-by-side outdoors, their not-so-glamorous backdrop was a barn. Julia's hair, then grey, parted in the middle and swept back off her face, lacked the curls of her earlier hairstyle. Her head tilted toward her husband, Julia wore little round wire-rimmed glasses above a pleasant, shy smile. The outfit she wore on her still-trim figure was more tailored and simple than her earlier one, yet still shows signs of planning.

In the black and white photo, Julia wore a simple skirt with a light-colored blouse trimmed with small, self-fabric ruffles on the front and cuffs. Her accessories included a dark necktie and belt typical of the early 1900s fashion era. John, a white-haired man with a white beard, looked at least sixty-nine because of his frown. In John's defense, his frown might be a squint against bright sunlight rather than an expression of his mindset. Dressed in a suit with a bow tie, John held a 1890s-era Homburg hat on his lap.

John played the fiddle; on weekdays, Julia approved. Julia was from a musical family and enjoyed his music. John also purchased a music box that he kept under an upstairs bed and brought out often to entertain children. On Sundays, however, John kept Julia busy trying to put away his music-making possessions. She considered them too raucous and desecrating for the Sabbath.

After their children were older, Julia and John moved closer to town. John, called *a walking newspaper*, was an avid reader who kept abreast of politics to discuss while visiting neighbors, the post office, and local stores. When I read this description of John, who died before I was born, I clearly understood his habits. They were the same traits I observed in his son, my grandfather. With a family to mother and household chores to accomplish, perhaps Julia read her Bible but had little time for recreational reading.

John's family was active politically. John's father, a staunch abolitionist, had been active in the Underground Railroad, and helped lead the early Republican movement in their county. John's younger brother, Joshua, served in North Carolina government, and helped establish training programs and schools for former slaves. Julia had married into a strongly opinionated and assertive family. Assertiveness wasn't new to Julia, whose own parents readily voiced and acted upon their convictions.

John and Julia ensured their children's education in keeping with Quaker tradition. We'll look at the lives of their three daughters later in this chapter. My grandfather attended nearby Guilford College, begun as a boarding school in 1837 by his father's ancestors and their contemporaries. Marvin's love for baseball surpassed his interest in scholastic endeavors. With or without their blessing, Marvin appears to have made his own choices. College baseball led to a lucrative career.

I've wondered if Julia's viewpoint might have been similar to the opinion of another Quaker woman in her neighborhood who had baseball-playing sons. Criticized when she let her sons, who were my grandfather's close friends and fellow ball players, play baseball even on the Sabbath, Julia's neighbor replied she did not worry about where her boys were or what they were doing. She knew where her boys were when they weren't at home; they were on the baseball field.

Julia's youngest daughter married first. On March 1, 1900, Mozelle, then fifteen but mature beyond her years, married. Mozelle's fiancé, Carl, was twenty-one, skilled, diligent, and responsible. Julia and John gave their blessing.

When the census taker visited John and Julia in June 1900, their household was missing not only Mozelle but Marvin as well. Marvin, then twenty-one, was living with his paternal grandfather, Andrew. Andrew, then a seventy-nine-year-old widower, was a farmer in 1900. He had previously earned his living as a merchant. Andrew's household included his son, Shubal (whose wife had died), two of Shubal's young sons, and Marvin who was recorded as a farm laborer. Marvin had attended Guilford College for three months earlier that year but his grandfather must have needed help with his farm during the summer.

John, who called himself a farmer, headed Julia's 1900 household. Julia was keeping house and her two remaining daughters, who had no official occupations, completed the listing for their family. Julia's oldest daughter, Mamie, was twenty years old and had attended school for eight months during the previous year. Lillian, Julia's third child, was eighteen, and had attended school for three months during the previous year. In addition to sending Marvin to Guilford College, Julia's single daughters continued their educations.

On February 6, 1903, Julia's first grandchild, Mary Kathleen, was born. Mozelle's first child, Kathleen became a brilliant career woman then followed Julia's example by marrying a man named John. Kathleen, who had no children, enjoyed mutual admiration with her nieces, nephews, and several generations of cousins. She died of cancer in 1991 at the age of eighty-eight.

Julia's second grandchild, Ralph, joined Mozelle and Carl's family on June 28, 1906. Mozelle gave birth to her third child, Julia "Judy" Ellen, on November 20, 1908. This baby's name must have thrilled her grandmother. Joy over Judy's birth quickly mingled with sadness. Perhaps busy with days-old infant Judy, Mozelle looked away for a moment and was horrified when toddler Ralph pulled a pot of boiling liquid off the stove, suffered severe burns, and died at the age of two on November 27, 1908.

Judy married and had two children while Julia was living. The only great-grandchildren Julia knew were Judy's son, Harvey (born in 1927), and daughter, Suzanne (born in 1930). Judy died in 1999 at the age of ninety-one.

Perhaps fearful she had too many children to manage, Mozelle waited six years before having her last child, Charles, on December 12, 1914. An evident Christian and faithful Methodist, a skilled engineer and military veteran, Charles admirably accomplished his many roles as husband, father, grandfather, great-grandfather, nephew, cousin, active church member, and neighbor. My encourager and editorial consultant, Cousin Martha, shares the privilege of calling that wonderful man *Daddy* with her brother, John. Charles lived a long, productive life and died in 2009 at the age of ninety-four.

Lillian inherited her maternal grandmother's dark hair and eyes, had a sweet disposition, and developed lovely manners. Mutual friends introduced her to Thomas, a young man attending college at the University of North Carolina. Thomas, quickly fell in love with the attractive southern girl; he proposed, and they were married in 1909. Thomas was Catholic. He and Lillian were married in a Protestant ceremony in her parents' home, and in a second ceremony performed by a Catholic priest. Julia and John welcomed and loved their son-in-law.

Tom's career took the couple to New York then to New Jersey. In 1910, Lillian gave birth to her first child, Mary Julia, later a pretty girl who charmed her southern cousins when visiting her grandparents. Appropriately named, Mary Julia inherited Julia's dark curly hair and lovely brown eyes. An avid reader, Mary Julia progressed well in school until she became ill with tuberculosis. After her parents doggedly sought the best medical help available, Mary Julia died in New York on December 10, 1928. Julia had lost her eighteen-year-old namesake granddaughter.

Lillian's second child, a son named for his father and called Tommy, was born in 1915. The family eventually returned to North Carolina; Tommy graduated from North Carolina State University, and then married Mary Ella, a widow with a young daughter. Tommy and Mary Ella built a home on his mother's family land, bequeathed to him by his aunt Mamie, where he lived until his death in 1994.

Julia's second daughter, Mamie, became a handsome woman with blue eyes, dark hair, and chiseled features. Mamie never married, continued to live at home, and cared for her parents until their deaths. She suffered for many years, uncomplaining, with rheumatoid arthritis, then died of cancer. Known as a woman of wit and integrity who was helpful to family and friends, Mamie continued as a faithful member of the Religious Society of Friends throughout her life. According to Kay Davis Coltrane's wonderful book, *Centre Friends, The Legacy of the Meeting on the Hill*, Mamie served as the first secretary of their women's missionary organization and served on a committee to improve music.

Marvin's love of playing and coaching baseball delayed his marriage. When he finally did marry in October 1911 at the age of thirty-two, he had a ready-made family. Lena, as you know from her story, was a young widow with a son. Since Marvin's bride had grown up in the same small community, his parents knew her personally or at least knew her by reputation.

I've brought Lena into her mother-in-law's story because the only negative I've learned about Julia came from the lips of her only daughter-in-law, Lena. During the early years of Marvin and Lena's marriage, they lived for a short time in Julia and John's household. Lena's first problem with the arrangement was, in her words, "too many women in one household." Julia's daughter, Mamie, was still living at home. When Lena moved in, there were three adult women in a small house. Lena said she found herself doing all the housework and cooking to her own (higher, in her mind) standards. Lena endured the situation for as long as she could, then she and Marvin moved their little family to her father's house.

We don't know Julia's side of the story so, in fairness, let's look at it from a different perspective. In Julia's defense, perhaps her homemaking shortcomings were a result of limited training and experience in her early years. By the time she was old enough to help, Julia's mother and older sisters were keeping house with assistance from hired help. Although not applauded for her homemaking skills, we admire her diligence at prayer and Bible study.

Julia lived to see the birth of Marvin's eight children mentioned in Grace and Lena's chapters, Lillian's two children, Mozelle's four children, and two of Mozelle's grandchildren (Julia's great-grandchildren). Although Julia and John lost two grandchildren, all four of their children outlived them.

While I knew all my grandfather's sisters, the one I knew best was his youngest sister, Mozelle, who lived until 1968. She was a wonderful person to whom we were all drawn and whose story bears repeating. Aunt Mozelle, as I called her, had a great twinkling smile and exemplified the joy of the Lord. Marvin and Mozelle had a close sibling bond, naturally happy temperaments, and were at ease other people; however, beyond those characteristics, they were quite different. As you will learn, Mozelle was industrious. My grandfather, you will recall from Lena's story, happily allowed others to accomplish necessary work because he preferred socializing to labor. Mozelle managed to do both with apparent ease.

Thanks to Mozelle's granddaughter, Martha, I've learned details about Mozelle's daily life that help define this youngest daughter of Julia. A great southern cook, Mozelle was widely known for her excellent food, including specialties such as persimmon pudding and pumpkin pie, prepared without consulting written recipes. In her later years, she opened a sandwich and soup lunch shop in the basement of a local school.

Cooks always need raw materials. When Mozelle was operating her lunch shop, her son, Charles, drove her to purchase supplies on Saturdays. When purchasing dairy products became costly and supplies were limited, she kept a cow and did the milking herself. She produced enough milk and butter for her own use as well as to supply several neighbor families. She raised hogs, made her own sausage, and hickory-smoked her own hams.

Mozelle was an avid gardener. She grew beans and other vegetables as well as pears, peaches, grapes and apples. She canned or carefully stored vegetables and fruit for the winter months. Her granddaughter, Martha, has said Mozelle never learned to sew because her two older sisters, Mamie and Lillian, took care of the family sewing. Apparently, the sisters helped one another by pooling their skills.

Mozelle's husband was equally talented and industrious. In addition to working with his wife to garden and raise meat, Carl was a skilled carpenter who constructed their first home.

As if the work I've described weren't enough, Mozelle found time to be an active Methodist. Her grateful church loved Mozelle for the many years she taught children's Sunday school classes. She sang in the church choir, yet that wasn't quite enough singing for Mozelle. She sang hymns as she worked. Her son remembered Mozelle singing every morning when he arose. What a great way to start the day and leave a sweet memory for your children.

When Mozelle passed away, someone wrote that she loved people and was always striving to see the good and beautiful in each person. I could see Mozelle's face in my mind's eye when the writer went on to say she always had a "merry word of cheer" to offer. Mozelle taught her children well so the excellent qualities of her life did not cease when she drew her last breath. Her descendants carry on her dedication to the Lord's work.

After Julia's mother died in 1886, Julia helped care for her father, Sampson, who lived nearby in his declining years. He walked with a cane in his later life, and carried a horn because his hearing was failing. An older granddaughter, raised by Sampson and Hannah when her parents died, lived with Sampson until his death. Toward the end of Sampson's life, Julia and her siblings hired a man to help nurse their father through illness and senility. Sampson died on September 13, 1895.

After fifty-four years of marriage, Julia's husband preceded her in death. John died of heart disease at the age of eighty-three in July 1932. After John died, Mamie lived with Julia, and Julia's granddaughter, Kathleen, settled the details of John's estate. John's 1933 will bequeathed to Julia land worth $4,250.00, household and kitchen furnishings worth $3.00, livestock worth $150.00, farm implements worth $25.00, and a car worth $40.00. At the time of John's death, their bank account balance was a meager $88.01. Obviously, amassing personal wealth had not been their goal.

When Julia died in August 1935, her death certificate indicated she died of a gallstone blocking her cystic duct. Julia's last few days of illness were marked with severe pain, fever, chills, and vomiting. Julia's remains were buried next to John's grave in the same Quaker cemetery where their parents were buried.

What did Julia's her life produce? She and John raised four children. When she died, all fifteen of Julia's grandchildren had been born, and she had three great-grandchildren. Julia's entire life had been lived within a small area of one state with no evidence she traveled beyond her local area. Did Julia's children have a positive impact on the world? We'll look at the two I knew best: her son, Marvin (my grandfather), and her youngest daughter, Mozelle.

If Julia influenced Mozelle's faith, work habits, and personality, both women deserve a round of applause. Mozelle was a singing, joyful daughter, wife, and mother, whose love for the Lord was apparent to all who knew her. Mozelle influenced her children and grandchildren toward strong relationships with God that became evident in their service to their churches and communities.

My grandfather, Marvin, faithfully attended church yet he failed to pass along spiritual strength to his children, many of whom struggled with alcoholism, depression, and other signs of lives adrift without the anchor of faith. He seemed to consider church an appealing social interaction rather than a time of reflection, confession, and edification. Thankfully, some of his descendants became people of faith and taught their children godly principals. Many of those who initially struggled with poor life choices later received the forgiveness and strength of God to turn their lives around for good. If Julia had known the struggles some of Marvin's descendants would face, she might have been more diligent about training Marvin. Some astute children learn by example; others need a more direct approach. As the Bible teaches, we must *train* future generations in the way they should go.

Julia's mother, Hannah, appears to have understood God's plan for at least one child and did her job with excellence. Hannah's teaching ministry bore fruit in Julia's younger brother, Eddy. Perhaps Hannah's prayers guided Eddy to his choice of vocation and prepared him to hear the Lord's call. Hannah did not live long enough to watch Eddy at work for the Lord, so we'll view his career as part of Julia's life and tell his story here. When Hannah died, perhaps Julia did more than watch her baby brother's career. It is possible she picked up where their mother left off and prayed for God's blessing on Eddy's ministry.

Around the time of Hannah's death in 1886, Eddy left home for Moody Bible Institute in Illinois to study for the ministry. In 1888, he began his work as a minister in the North Carolina Methodist Conference. Julia was alive for all the years of his ministry and well beyond his retirement, so she knew of his accomplishments first-hand.

Julia was available not only to applaud Eddy's triumphs but also to comfort him during his sorrows and family challenges. He married, his first wife died, and he married again. Although neither marriage produced children, Eddy's nest wasn't empty. He and his second wife raised his niece, Helena, after her parents passed away. Helena was the daughter of Eddy and Julia's eldest sister, Virginia, who passed away in 1881 after being widowed in 1880.

From a pastoral career standpoint, Eddy experienced the frequent church-to-church moves so typical of Methodist ministers of his day. Although not assigned to a church close to Julia's home, he traveled as an evangelist for twelve years and could have preached revivals or camp meetings close to Julia's home. Perhaps Eddy also performed baptisms, weddings, and funerals for family members who followed in his Methodist form of worship. He retired in 1923.

After Eddy died in early 1946, a tribute to his ministry mentioned his gift at preaching revivals. His peers estimated his sermons prompted thousands of people toward birth into the kingdom of God. They attributed Eddy's success as an evangelist to his winsome telling of the story of Jesus and His love.

The results of Eddie's career must have pleased Julia if they reached her ears. During one revival meeting altar call, over one hundred people of all ages responded. They came forward seeking salvation, baptism, and church membership. The pastor who penned Eddy's tribute wrote that he was a lad in his early teens and was among those who came to know Christ that night. Apparently, as Eddy spoke, he told the group he believed God was calling one of those boys to be a preacher. The writer explained that as the years passed, he realized he was the individual called to preach that night, and ultimately began his own ministry two years after Eddy's retirement. The writer believed through his own ministry, Eddy's love and labors continued to bear fruit.

I wonder if the prayers Julia prayed under the stairs included prayers for the success of her brother's ministry. If so, perhaps her prayers blessed not only people who heard Eddy, but also blessed those who heard men Eddy inspired to enter the ministry. How far-reaching were her prayers? Her prayers could have a never-ending lifespan. Imagine that.

The accounts of the lives of others among Julia's eight siblings appear in the story of her mother, Hannah, later in this book. Julia's youngest brother, Eddy the minister, outlived her; however, Julia's other siblings predeceased her. She had apparently been close to her older brother, Robert, who moved to Illinois after the American Civil War. After his death, his wife told family members that he cherished his little sister, Julia, and kept a photograph of her nearby all his life. Robert passed away in Illinois.

Julia's brother, William, died in Illinois as well. In 1911, her brother, John, died in Texas where he had raised his family. Once her three eldest brothers moved west, she probably never saw them again. Julia was only thirty-two when her eldest sister, Virginia, died in 1881 followed by her sister Hattie in 1928 and Martisha in 1931. Perhaps the sibling with whom she spent the most time was her brother, Frank, who died in 1910. Frank, two years younger than Julia, married Martha, a sister of Julia's husband, John.

Frank and Martha (who married six years before Julia and John) lived close to Julia while they raised a large family of ten children. Frank and Martha's children were double first cousins of Julia and John's children. When Frank died in 1910, Julia must have missed him terribly. Frank's wife, Martha, then moved to Richmond, Virginia, where several of her children had found jobs.

What might we learn from Julia's example? Have we underlined passages in our Bibles? If so, have we merely read them, maybe even memorized them, yet failed to make certain the generations after us know the Bible is a source of strength? Are our children prepared for this life *and* the next? Are *all* our children (even the difficult ones) spiritually equipped? Julia lived to be eighty-six years old. Do we have that much time? Don't wait. Start today.

Great-grandmother Julia was alive when two of her granddaughters and one niece became her namesakes. Since her death, one great-granddaughter and two great, great-granddaughters inherited Julia's name. If I could whisper in our great-grandmother's ear, I'd tell her about the Julia I've known best during my own lifetime: my first cousin, Julia.

Sixteen years after great-grandmother Julia died, her son Marvin's youngest daughter, Carol, gave birth to her first daughter. Born on February 26, 1952, Cousin Julia celebrates the anniversary of her own birth one day after the February 25 birth date of great-grandmother Julia. Cousin Julia also remained in Guilford County to marry and raise her children. As did her great-grandmother, Cousin Julia has worshiped in Quaker meetings and Methodist churches. Unlike her great-grandmother, I don't believe Cousin Julia goes into a closet under her stairs to pray, although Cousin Julia takes her dedicated prayer life to those in need and God blesses her efforts.

Cousin Julia is a prayer warrior. She has organized the cousins of our generation into a responsive prayer team, and keeps us informed of family prayer needs via e-mail. Thanks to Cousin Julia's leadership, we've seen awesome results of God's gracious intervention. Cousin Julia brings honor to our great-grandmother's name and faith.

Although not namesakes, Julia has many other male and female descendants who work tirelessly to honor God and support His work. Her descendants serve as deacons, elders, and teachers in their churches in North Carolina and other states around the country. Among Julia's great-grandchildren are two men dedicated to ministry that facilitates spiritual healing and deliverance from substance addiction.

Have all her descendants lived perfect lives? There are no perfect humans. Some of us have (praise God) avoided grave difficulties; others have fallen under the weight of life's trials and had to climb back into the light. God has been so faithful to meet and heal us right where we are when we turn to Him, regardless of our past. Did Julia mention future generations as she spoke to God in her little prayer closet? If so, we (whose names Julia never knew) continue to benefit from her spiritual strength today.

Let's all follow the lead of women in our ancestry who left evidence of their devotion to prayer and Bible study. If we don't find that evidence in our ancestors, let's become the first generation of many spiritually strong women. Whether your closet is literal or figurative, set aside time to pray. We know Julia spoke to God and we hope she took time to listen as well. He has the best plan for our lives. Remember the meaning of Julia's middle name: God is good.

Mamie

1868 – 1944

Two men perch between my twig and Mamie's branch on our family tree. Mamie gave birth to my father's father. Because my father said little about his father's family, half a century of my life had passed before I learned Mamie's name. Our estrangement from Mamie and her story began when the younger of Mamie's two sons, George, disappeared in 1921 when his son, John, was sixteen months old. John's paternal grandparents failed to cultivate a relationship with their only grandchild or to fill their absent son's role in John's life. John responded with equally little interest in them.

Society's positive view of Mamie and her husband makes Mamie's unbecoming behavior toward John and other family members puzzling. The inside story of Mamie's personality, particularly in later life, remains shrouded in mystery. In an attempt to understand Mamie and seek evidence of her spiritual condition, we'll review her busy, otherwise successful life. As you read about Mamie, two questions beg consideration:

(1) If Mamie's son voluntarily disappeared, was there anything she or her husband could have done during George's childhood or early adulthood to prevent his abandonment of his wife and child?

(2) Could Mamie and her husband have helped their grandson, John, become a spiritually and emotionally stable man rather than a violent alcoholic?

Even when parents dedicate their lives toward raising Godly children of sound character, children can inexplicably go wrong. Excellent grandparents can fail to see signs of trouble, feel it is not their place to become involved with grandchildren's lives, or become overwhelmed. Perhaps in Mamie's life, she was busy doing worthy work, yet failing to see signs of distress in George. Perhaps Mamie did everything possible and George chose to go his own way.

Mamie was fifty-two when her grandson, John, was born in March 1920. She would have been fifty-six or older in John's boyhood memories. John described Mamie as cold, stiff, and Victorian. During rare visits to her home, he recalled Mamie thrusting a plate of cookies toward him saying, "Here, have a cookie." When pressed, he knew nothing of Mamie's life and recalled no actual conversations with her. Failure to be present in her grandson's life precludes any possibility Mamie could influence his life for good.

Initially believing there was simply a relationship issue between Mamie and her son and, by association, her grandson, I didn't expect to hear my father's description of remote Victorian stiffness repeated by others. When I interviewed Cousin Emily, who knew Mamie when she was a child, she described Mamie as a cold, unfriendly woman, as did cousins who repeated comments made by their parents. In Mamie's defense, Cousin Emily was a young child when Mamie was an elderly woman. Emily and others whose opinions I've gathered knew Mamie only during the years following the traumatic event of her son's disappearance. As a child or younger woman, perhaps Mamie was a more pleasant, caring person.

It might sound as though we're getting ahead of ourselves by viewing Mamie as an older adult before we talk about her childhood and marriage. Knowing her reputation among close relatives in her later life informs our research of earlier years. What was the root cause of her apparent unhappiness? Was her personality an outward indication of a physical or mental illness? Was her apparent faith only window dressing? Did she lack an emotional support system? Was she overwhelmingly unhappy or depressed only in her later life? Mamie faced challenges; but when dark days come, we can summon strength from our relationship with God. Of course, to do that, we must first *have* a relationship with God.

To gain peace of mind before writing Mamie's story, I opened my Bible. When I read 2 Corinthians 8:18 I was certain God had directed my reading. That verse mentions a preacher of the good news. When such a man entered Mamie's life, she repeatedly heard the good news of God's plan for her salvation. I pray she responded by accepting salvation, even though she apparently allowed life's issues to stifle her outward joy.

Mamie's story can remind us to look for redeeming qualities. No human is perfect. Even people who leave a negative first impression usually have redeeming qualities. Our own happiness and peace rest heavily on our ability to find the good in people then focus on those positive traits. Negative traits might even disappear when we encourage others to bring their good qualities to the fore. Occasionally, no matter how hard we try, some people won't change. We're responsible for doing our part then leaving the results to God. In the process of seeing and fertilizing the best in others, we learn and grow ourselves.

Aside from a few early references where this woman's name appears as Mary M., her written and spoken name was Mamie. Her first name, Mary, honors her great-grandmother or aunt. Seen only as an initial, her middle name could honor her maternal grandmother, Margaret (the diminutive of which is Mamie), or her father's sister, Mary Matilda. Her only grandchild called her Mama Mame. We'll refer to her as Mamie in this chapter.

Now I'll begin at the beginning and tell Mamie's chronological life story, then we'll look at the sum of her life and legacy.

Mamie and I share a common birthplace: New York City. Her parents lived at 799 Greenwich Street when she was born in August 1868. Mamie's first home was six miles from my first home in the New York borough of Queens. Our birth experiences further parallel because we were born in New York City to southern parents who were there only temporarily for reasons pertaining to our fathers' careers.

Mamie drew her first breaths in a bustling northern metropolis, yet the voices surrounding her cradle lilted with a distinctive Eastern Shore of Virginia accent. Her father had taken his family north to market produce grown in their home county. Although loyal Virginians, her father and his family expressed abolitionist leanings during and after the American Civil War, so they probably felt politically in harmony with their northern neighbors. On the day Mamie was born, newspaper reports covered reconstruction issues and the story of women in Atlanta resolving to erect a monument to the Confederate dead. The nation was trying to move on, yet people wanted to remember what they had lost.

Let's enter Mamie's New York City home and meet her parents. Her father, Littleton Thomas, was called Tom in an effort to avoid confusion with his grandfather and an uncle who bore the name Littleton. We'll call him Tom in this chapter; Mamie called him Papa, as did all his children and grandchildren.

Mamie's mother was Cordelia Elizabeth; known as Mama to her children, and affectionately called Cordie by others. A detailed examination of Cordie's life appears in a later chapter of this book where facts about Mamie's maternal ancestors are included. In this chapter, we'll briefly explore stories of Tom's family as background for Mamie's life.

Tom's ancestors arrived on the Eastern Shore of Virginia in the 1600s. Tom's maternal ancestors were farmers, while early men of Tom's direct male line were sailors involved in West Indies trade. When Tom's grandfather, Littleton, sailed into Wachapreague Inlet, he met Esther who had inherited part of her father's land (from a 1662 ancestral land grant). When he married Esther, Littleton traded his life at sea for farming his wife's land and raising a family. Described as tenderhearted and a model housekeeper who was especially kind to servants, Esther had four sons and spent her life on the family plantation. Littleton never lost his love for the sea. His devotion to his family and his fondness for their home kept him from roving.

Family records describe Littleton as Esther knew him. A man with black hair and blue eyes, he was jovial and loved stories. Although not highly educated, he was a self-taught, excellent navigator. As a farmer, he became a successful, well-regarded citizen of the Eastern Shore. Prior to his 1837 death, Littleton requested the funeral text: *If a man dies, will he live again?* Perhaps Littleton and Esther provided religious training; one of their sons became a Methodist minister.

Littleton and Esther's eldest son (later Tom's father), Nathaniel, was born in Accomack County, Virginia, in 1808. Nathaniel had light auburn hair, blue eyes, and clear-cut, handsome features. He mastered a wide variety of skills enabling him to perform the work of a plasterer, brick-mason, blacksmith, weaver, and tailor. His sons reported their father cut out all the top clothes they wore during their childhood. Nathaniel also served as justice of the peace for his county as well as postmaster in his local area.

As the eldest son, Nathaniel inherited Esther's ancestral land. After his father's death, Nathaniel moved an older section of Esther's ancestral home to a new site, and then spent eight years adding a larger addition to accommodate his growing family. The resulting house, known as Atlantic View, literally afforded a view of the Atlantic Ocean. Mamie and other members of her father's family formed an emotional attachment to Atlantic View, where Tom and his siblings spent their childhood and the children of Mamie's generation visited often. Tom's brother, Nat, was a noted author who immortalized their childhood home in a lengthy poem that begins:

Atlantic View, my boyhood home, down close beside the restless sea;
No matter where my footsteps roam, my longing heart still turns to thee.

If men learn to be a good husband and father by watching their own father, Tom had a great example in Nathaniel, described as active, quiet, cheerful, a most faithful husband without reproach, and a kind, generous father. His son wrote that Nathaniel's shortcomings were a lack of understanding of human nature and a limited imagination. Mamie knew Nathaniel well but his wife, Elizabeth, died when Mamie was two months old.

After his first wife died in 1868, Nathaniel lived in the North with his sons and their families for several years before returning to Virginia's Eastern Shore and remarrying. In 1875, he married Julia, a woman twenty years his junior, who outlived him. At his death, Nathaniel left the ancestral family home, Atlantic View, to Mamie's parents to enjoy then pass on to their children.

Littleton Jr. (Mamie's great uncle) was a spiritually noteworthy character. Facts about his life provide a glimpse of religious choices among Mamie's family members in her grandfather's generation. Called into ministry in his early twenties, Littleton Jr. served as a Methodist minister for fifty years. Apparently, to distinguish him from his father, he used his initials, L. K., as his call name. Rev. L. K. was a man whose naturally kind, sympathetic manner made him welcome in the sickrooms of his flock.

Rev. L. K. might have directly influenced Mamie's life because he related well to children. The story of his life mentioned that long after an attack of paralysis ended his active service to the church, he continued talking to children when he was physically able. He lived until January 2, 1882, when Mamie was thirteen. Even if she had not spent much time with him, she knew the details of his ministry. During a visit to Trinity Methodist Church in Salisbury, Maryland, seeking information about a later pastor who served there (Mamie's husband), I learned Rev. L. K. was the first minister of that church when it was formed in 1866.

Having introduced Mamie's paternal ancestors, we'll return to her early childhood. When she was born in New York City, her parents had welcomed five children in Accomack County, Virginia. One little brother passed away before Mamie was born, so Mamie joined a sibling group of two brothers and two sisters. Thanks to a December 1868 letter written by Mamie's mother, I learned "little Mary" (who wasn't yet called Mamie) was described as "good, playful, and a general favorite" when she was four months old.

Mamie soon relinquished the spotlight to a new infant. Susie, a sister with whom Mamie remained close all her life, was born in 1871, followed by Eva in 1874, and Emma in 1877. The latter two sisters were born in Baltimore placing Mamie there when she was six. The family also lived in Pennsylvania and New Jersey, although the precise dates of all their moves remain elusive.

In December 1880, Mamie's parents left for Jacksonville, Florida, hoping the change of climate would improve her mother's health. Cordie was ill with consumption. Although they must have known death was a possibility, I don't know whether Mamie (who was twelve years old at the time) realized she would never see her mother again. The only child permitted to accompany their parents to Florida was Mamie's youngest sister, Emma, who was also exhibiting signs of illness. Cordie, forty-two, died on January 11, 1881, leaving Tom with eight children ranging in age from twenty-two to three. Mamie had lost her mother.

Probably because he felt ill equipped to raise so many children alone, Tom didn't remain single. Later in 1881, he married Sallie, a thirty-year-old North Carolinian. I have found no direct opinion from Mamie with respect to her stepmother; however, her youngest sister, Emma, said Sallie was a great mother.

A June 1884 letter written in Baltimore by Mamie's younger sister, Susie, mentioned Mamie (then fifteen) and her sisters were awaiting results of their school exams. By August 1884 the family moved to Camden, New Jersey. Mamie's half brother, Thomas, was born in Camden soon after their arrival. Thomas died at the age of two months, two days. Perhaps the move had proved too strenuous for Sallie.

While they were living in Camden, Mamie's older sister, Peddie, died on Saturday, January 17, 1885, at the age of nineteen. Mamie's father recorded Peddie's burial in the family Bible where his note reads, "We gently laid her remains next to those of her dear mother's, whose gain is our loss." Tom went on to write that Peddie was "devoted to her church and dearly loved by all her associates."

Mamie was seventeen, and old enough to help with an infant, when her stepmother gave birth to a daughter. Born on January 8, 1886, baby Sallie (named for her mother) became an accomplished musician and music teacher, never married, lived until 1971, and was reportedly well-loved by her half-siblings.

In May 1887 Mamie and her family again lived in Baltimore and made summer visits to Virginia. In a note of that era written (partly in French) by Mamie's younger sisters to their cousins, a date was made to meet on the croquet lawn. Apparently summer life included times of play among members of Mamie's extended family.

Tom willed Atlantic View to the children of Mamie's generation. Mamie's sister, Emma, later purchased Atlantic View from her siblings and raised her family there. Emma's daughter, Cousin Emily, was born in 1917 and lived in the home during her childhood. She graciously provided a verbal description of the interior.

The first floor, according to Emily, consisted of a hallway, the dining room and a parlor. The first level also included a porch to allow the family to enjoy summer breezes off the ocean. There were four bedrooms on the second floor. The kitchen, in Emily's childhood, was still in a separate building. There was no electricity or indoor plumbing. Cousin Emily was thankful for the installation of plumbing and electricity during her later childhood. Mamie, who visited Atlantic View before and after the addition of those conveniences, must have agreed.

Cousin Emily, who was present during Mamie's later visits, said family meals included seafood or meats such as fish, clams, crabs, ham, pork loin, pork chops, and sausage. They filled out their meals with locally grown vegetables such as turnip greens, cabbage, kale, butter beans, corn, and onions. Emily remembers a family tradition of making yeast rolls weekly. A family dessert favorite was a chocolate cake created by filling four thin layers of cake with lots of icing. They enjoyed pies made from blackberries and huckleberries picked from bushes along the sides of the roads. Manners, according to Emily, were always important at the family table.

Mamie attended a Methodist camp meeting in the summer of 1888 that changed the course of her life. The guest pastor leading the camp meeting, Brother John, as Mamie's family later called him, was stationed at Trinity Methodist Church (the church Mamie's great uncle pastored beginning in 1866) in Salisbury, Maryland. John looked out into the congregation, Mamie caught his eye, and John reportedly said, "That is the woman I'm going to marry."

Something about John must have pleased Mamie as well because they were married a few months later on Christmas Day 1888. Mamie, twenty, became the wife of a twenty-six-year-old Methodist minister. What did they see in each other? Physical attraction is easy to imagine since a photograph taken at the time of their marriage capture John, a handsome man with a handlebar mustache, and Mamie, attractive with light eyes and a serene expression.

Cousin Emily described elderly Mamie as a tall, slender woman with straight, stiff posture. She had medium brown hair that was graying by the time Emily knew her. Her eyes were light but not impressive according to Emily who guessed them to be grey or an unremarkable shade of blue. In photographs taken after her marriage, Mamie's clothing and hairstyles appear appropriately simple and understated for her role as a preacher's wife.

Perhaps Mamie inherited her family's love of literature and impressed John as a well-read conversationalist during their courtship. Mamie, though not a striking beauty, had a look of attractive intelligence. As part of Mamie's story, we'll explore John's paternal ancestral line. You'll learn the story of his maternal ancestors in the chapter about his mother, Annie.

John was born in Somerset County, on the Eastern Shore of Maryland. Since the 1660's when John's ancestor, William, moved to a land grand in Somerset County, John and his paternal ancestors had been born within a small area of that county. William, our Dutch ancestor who came to America by way of England, had first lived in Northampton County, Virginia beginning about 1647 and probably crossed paths with Mamie's early Virginia ancestors. John's family, like Mamie's, arrived early, becoming the first Europeans to own land later willed to their descendants.

Let's skip directly to John's father, Isaac Henry, who began his working life as a sailor. He became a ship Captain, owned a fleet of fishing and shipping boats, and later operated a fertilizer factory. I don't know whether John was close to Isaac, or whether Isaac was a good father-in-law to Mamie. John preached one sermon that might indicate Isaac was a man of faith. A Richmond newspaper reported *My Father's God* as one of John's sermon topics in January 1900. I'd love to know what Mamie heard John preach that day.

During the early years of John's life, his family lived with or near his father's widowed mother in Somerset County, Maryland near Princess Anne. By the time John was seven years old, his father supervised a fertilizer business called guano works, and moved his family to Norfolk, Virginia to be close to the location of his business. He and Annie raised their boys as Methodists. John attended public and private schools in Norfolk then studied for the ministry at Virginia's Randolph-Macon College.

Trinity Methodist Church of Salisbury, Maryland, his church when he married Mamie, was his third charge. First assigned to a church in Norfolk, Virginia, John then became a domestic missionary to Newport News, Virginia. Having begun his ministry as a single man, he was probably happy to have a wife to help with the spiritual and social aspects of his work.

As I previously mentioned, I was eager to visit the first church Mamie joined as a married woman. Initially disappointed the current building was not the structure that existed when John first pastored there, the visit proved quite fruitful. The gracious people of Trinity, a vital, active congregation when I visited in 2007, provided copies of John's handwritten reports to his Methodist superiors.

In John's report of February 14, 1889, Mamie's name appears among a group of four new members received into membership by certificate. John's bride of seven weeks, Mamie's membership by certificate indicated she had been a Methodist church member elsewhere. Reviewing other notations in John's pastoral reports, I've gained a sense of his spiritual maturity.

John reported the people of Trinity were praying for a revival, and praying for a gracious outpouring of the Holy Spirit. He noted the congregation was growing, and more people were receiving the Lord's Supper regularly. John may have been looking forward to being the father of sons when he wrote, "I, as pastor, have a class of bright boys whom I am trying to instruct and bring to Christ."

John's reports mention holding quarterly meetings of church leadership at the parsonage. Official attendees of those meetings included John (as pastor), a presiding elder, and the secretary of the church. At the least, a meeting in the parsonage meant Mamie's home had to be presentable, and perhaps she served refreshments. Although the pastor's wife was not a named attendee, I wonder if Mamie joined the meetings or discussed the proceedings with John when the others had gone home.

Mamie was pregnant when John made his May 1889 report about guiding the births of new spiritual children. In addition to five new members received by certificate, John reported twelve new members received on profession of faith. His ministry was bearing fruit. Perhaps Mamie was praying for him and encouraging him. On October 25, 1889, John reported having baptized six infants. Records for the period indicate John's salary, $226 to $240 per quarter, was dependant upon money collected from the congregation and other church expenses.

John made three to four times more money in his city church than did my other great-grandfather who pastored smaller churches during the same era. John's higher salary speaks well of his congregation's giving habits, or perhaps reflects the financial stability of a long-established church. Whatever the reason, it appears Mamie had enough financial support as she began her married life and started her family.

Mamie gave birth to her first child, Robert, on December 3, 1889, in Salisbury. Named Robert to honor John's maternal grandfather and older brother, Mamie's family surname became his middle name; a naming convention she continued with her second son as well.

By the middle of 1890, while Mamie was busy caring for her first child, John's reports mention continued prayers of his congregation. "Some of us," he wrote, "believe that in the near future, God will visit this part of his course with a genuine revival of religion. For this, we are praying and hoping." During the same period, thirteen people joined their church by certificate, three adults joined by profession of faith and baptism, and John baptized two infants. John reported the need for more Sunday school teachers. He soon reported having four teachers prepared and willing to teach whenever called upon.

In November 1890, John continued to report the addition of members by baptism and certificate, and seemed unable to contain his pleasure at how well the work was progressing. He described all services as "well attended" then mentioned the Wednesday night service being "a source of great joy and power."

An 1890 Methodist publication described John as a man of average stature and weight, with an open countenance that "admits him at once into the confidence of all people." The article went on to say John was never at a loss for ideas, exhibited brilliant imagination in an engaging style, and spoke from memory in a clear, penetrating voice. Mentioning John and Mamie's 1888 marriage, the article went on to say they were both popular and made friends wherever they went. Mamie was becoming an asset to her husband.

Sallie, the second wife of Mamie's father, died on January 3, 1891, four days before her little daughter's fifth birthday. Her stepmother's cause of death, tuberculosis, was sadly familiar to Mamie and her family. Mamie's mother had died of TB, leading us to speculate her father unwittingly infected his second wife. Since Sallie had been kind to Mamie and her siblings, her death must have been difficult for them to endure. Mamie's father had lost another beloved wife. In a note to their little daughter on a family keepsake, Tom wrote, "I am in deep trouble on the loss of your sweet Mama." Many years later when Tom died, one of his friends commented that Tom met the experiences of life with remarkable composure so he must, at least outwardly, have appeared to soldier on following the loss of his second wife.

In accordance with their practice of moving pastors about every two years, the Methodists sent John to pastor a church in Culpeper, Virginia in 1891. Leaving the Eastern Shore for the first time since her marriage, Mamie might have felt apprehensive about the separation from her extended family; yet the many relocations of her childhood had prepared Mamie for the nomadic life of a preacher's wife.

Mamie established her second household in the historic town initially called Culpeper Court House when the first post office opened there in 1795. Steeped in American Revolutionary and American Civil War history, Culpeper offered a stimulating change if Mamie had time to enjoy the stories and see the landmarks while she raised her baby, Robert, and continued life as a busy pastor's wife.

Mamie was seven months pregnant with her second child when her brother, Edwin, died in Chicago on October 19, 1891. Although sent for burial among his father's ancestors in Quinby, Virginia, I believe Mamie's advanced pregnancy precluded her attendance at Edwin's funeral and burial.

On Christmas Eve 1891, Mamie gave birth to her second and last child whom they named George. The source of my grandfather's first name could have been one of John's brothers or Mamie's sibling, uncle, or grandfather. In keeping with tradition, George's middle name was Mamie's maiden name.

By 1893, Mamie's father was married to his third wife. At the age of sixty-two, Tom married his thirty-nine-year-old maiden cousin, Sudie. Although Sudie outlived Tom, when she died in 1915 the cause was again tuberculosis. From Tom's perspective, he had Sudie's help to raise the daughter born during his second marriage and companionship in his later years. Again, as with Tom's second wife, Mamie and her siblings reportedly loved their cousin in her role as their stepmother.

In 1895, John transferred to Wright Memorial Methodist Church in Portsmouth, Virginia. We know they made a trip to the Eastern Shore at least once in 1895. When Mamie's younger sister, Eva, married in April, Mamie and John were there and John performed the ceremony "in a beautiful and impressive manner," according to the newspaper account of the wedding.

On May 15, 1896, John's sixty-six-year-old father, Isaac, died of liver cancer in Norfolk, Virginia. Because Mamie and John did not live in Norfolk during Isaac's lifetime, Mamie wouldn't have known Isaac well. She and John attended Isaac's funeral, and then probably spent time with John's brothers planning how best to comfort and assist John's widowed mother, Annie.

In 1897, John and Mamie received their western-most assignment. They moved to pastor beautiful Mount Vernon Methodist Church in Danville,

Virginia. Lin, a kind Secretary of that church, graciously provided facts about John's time there. She mentioned John's salary during the period was $1,100 per year plus $300 for the rent of the parsonage. A book of church history, *Sixty Years at Mount Vernon,* mentions controversy about a new suit of clothes for John. Someone suggested the church purchase a suit for him. Some church leaders opposed the idea, and deemed it more appropriate to raise their pastor's salary so he could purchase the clothes he needed. If John's clothes were threadbare, he and Mamie must have had little money to cover their household expenses, purchase Mamie's clothing, and meet their children's needs.

During their time in Danville, records indicate John conducted Epworth League meetings, and was interested in collecting funds for a home for the sick. A church member described John as an optimist who could find grounds for hope in all circumstances. John's Danville congregation attempted to stop his removal to Richmond at the end of his prescribed stay.

John preached in the Mount Vernon sanctuary still in use when I visited that beautiful church in September 2010. Because of my interest in Mamie, I stood in their sanctuary and imagined her sitting there listening to her husband. Had she already heard his sermon as he prepared it, or was it new information to her? Did she pray for him as he preached?

Mamie and John were in Danville on June 11, 1897, when Mamie's maternal grandmother, Margaret, died in Accomack County, Virginia. If they could get away and had money to travel, I'm sure John and Mamie joined her family for the funeral. Margaret welcomed Mamie and her siblings into her home many times; most notably, she cared for them when their mother was dying. Margaret was a strong woman who blessed her children's children.

John's 1899 salary level was more meaningful when I read newspaper ads of the era. John could purchase a two-piece suit for $10.00 to $30.00 and a dress shirt for as little as fifty cents. Mamie wore skirts with fitted tops called *waists.* In 1899, Mamie could buy a skirt for thirty-nine cents.

Beyond clothing, Mamie was stretching John's $1,100 per year salary to cover household expenses. In 1899, shredded wheat cereal cost fifteen cents a box, soup beans were eight cents a quart, and bologna was ten cents a pound. Other household purchases might have included quilts at a cost of forty-eight cents each.

A newspaper article published on April 1, 1899, announced John's assignment to Hallwood Methodist Church in Salisbury, Maryland, the town where they had begun their married life. Perhaps John was serving in an interim capacity because they didn't stay long. By December 1899 they had moved to Richmond.

In Richmond, John pastored Park Place Methodist Church (later called Pace Memorial and now a student center). They lived nearby at 613 Grace Street. A letter written by Mamie's father to her sister, Eva, indicates Tom is seeing Mamie's Richmond home for the first time and implies they haven't been there long. Tom's words provide a bit of insight about how Mamie and John were fairing in their Richmond assignment.

December 11, 1899

Dear Eva,

I write to say that I had a pleasant trip over. I found our folks all in good shape. Went to church morning and evening. Had a full house and everyone seemed pleased with John. Mamie now is in her glory but these occasions tend to wear her out. Their house is fully furnished in every way and they only have to meet the demands of their church people. It is an easy walk over.

Love to all,
LT

In an undated note from the era, Tom wrote, "Mamie comes over next Tuesday. Bro. John is to carry on a revival at Cape Charles, and she is coming over to see Eva and me. We are all to be at Eva's next Saturday for the night." From the wording of Tom's note, Mamie's family tried to gather for a visit when Mamie and John were on the Eastern Shore. Mamie must have enjoyed those times with her siblings. What she couldn't know was how little time she had left with some of them.

Numerous newspaper articles documented John and Mamie's first full year of ministry in Richmond. I was astonished at the number of activities

John supported in addition to work at his assigned church and pleased to learn of Mamie's activities as well.

I've summarized the highlights of John's numerous activities. In 1900 alone, John participated in twenty-two events important enough for newspaper mention. The twenty-two events did not include his preaching schedule on Sunday mornings and evenings, or the weddings, funerals, and the impromptu meetings that fill every pastor's weeks. John appears to have little time for his family, so Mamie must have been busy keeping house, raising their sons, and fulfilling her supporting role as the pastor's wife.

John was a member of a Masonic Lodge where he served as chaplain. He spoke at teacher's conferences, the Young Men's Christian Association, Sunday school rallies, missionary conferences, camp meetings, and other special services at various churches. He judged an orator's contest and served on the board of managers for the Bible Society of Virginia. He was apparently willing to serve in a leadership capacity in whatever organization he joined.

While Mamie was running a household and raising her boys, she led the Epworth League meeting at their church in January 1900 after they had only been in Richmond a few weeks. Named for the boyhood home of the founder of Methodism, John Wesley, the organization of Methodist young people began in 1889. Mamie probably led the group to remember their motto of *Look Up-Lift Up* during a program focused on one of their six stated activities in support of Methodist work: spiritual life, social work, literary work, correspondence, mercy and help, and finance. Epworth League members pledged to love their fellow men and women for Christ's sake, uphold a standard of moral purity, and help further the cause of the temperance movement.

In a photograph of the era, an Epworth League pin decorates the center of Mamie's collar, so perhaps the organization was important to her. Since her husband often led the meetings, I don't know how comfortable she felt when asked to take a leadership role so early in their Richmond ministry. Whether it was difficult for her or not, she accomplished her assignment. Later in 1900, John led the committee on public worship at a convention of the Virginia Epworth League. They were apparently a team.

Mamie joined the Women's Club of Richmond during the 1900 timeframe. Populated by Richmond's leading women, the Women's Club membership records for 1900 don't survive. Fortunately, their meetings were society column news. The newspaper report of the Club's March 1900 meeting mentioned Mamie's mother-in-law and described her as a "guest from afar." I assume Mamie was a member and Annie accompanied her daughter-in-law to the meeting during a Richmond visit.

The 1900 census documented Mamie's Richmond household on June 6, 1900, as including John (thirty-seven), Mamie (thirty-one), Robert (ten), and George (eight). For Mamie's sake, I was happy to note an additional person in their home. Annie, an African-American young lady from North Carolina, was living with them and performing general housework. I'm sure Mamie appreciated housekeeping assistance while she busily met her social and church obligations.

City life in Richmond offered advantages for Mamie. She could shop downtown and travel by electric streetcar. Another advantage of living in a city with growing boys was the availability of better schools. There were new educational venues such as the Museum of the Confederacy (opened in 1896) and the Valentine Museum (opened in 1898). John and Mamie could explain roles their family members played in the American Civil War and Virginia history while touring the exhibits.

On July 5, 1900, a Richmond newspaper item mentioned the illness of Mamie's eldest sister, Bessie, and that Mamie had gone to be with Bessie and her family. Six days later, the paper reported Mamie's sister was somewhat worse. Bessie died on July 16, 1900. Bessie left a husband and six children in addition to her father, siblings, and friends to grieve her premature passing at the age of forty-one. One of Bessie's children, a little girl, had been born in May 1900 and died two days after her mother.

Ironically, not long before her death, Bessie delivered a speech she had written for a gathering of Methodist women at her church. A phrase she had used, "may we each be faithful until we hear death's call," must have seemed particularly poignant to her family after her passing. A decade older than Mamie, Bessie comforted and helped raise Mamie and the other younger siblings when their mother died. Bessie's passing must have been difficult for Mamie to bear. She and her sisters spent that summer on the Eastern Shore to help Bessie's children through the difficult weeks following the loss of their mother.

Also in 1900, John's family had a reunion in New York City where his brother, William, and widowed mother, Annie, were living. I have wondered what Mamie's thoughts were upon returning to the city of her birth. By the time she returned, Miss Liberty had arrived to grace the harbor, and bridges spanned the rivers to facilitate travel to the outlying boroughs. I wonder if she visited the addresses where her family had lived.

John's busy schedule continued even with illness and death in Mamie's family. He preached for special meetings in King William County, Virginia, during the week of July 21, 1900; and joined a team of pastors at a camp meeting in Madison County, Virginia, beginning on August 3, 1900. John also switched pulpits with other pastors on occasional Sundays. From Mamie's perspective, I wondered if she went with him or if she had obligations at their own church. Either way, I'm sure John's busy schedule meant Mamie was busy as well.

Mamie's efforts to guide and encourage her own children continued no matter how busy she was with other activities. Her son George's name appeared in November 1900 among students who were achieving honors for their schoolwork.

In December 1900, John conducted a double wedding in their parsonage. Even though the announcement stated only relatives and intimate friends were present, I'm sure Mamie had to make certain her home was ready for the event.

Apparently, Mamie's busy husband traversed Richmond on a bicycle. On March 14, 1901, John was involved in an accident when two ladies driving a buggy hit his bike and threw him to the pavement. John sustained a back injury and a broken rib. After treatment by a physician, he went home to recuperate. Another pastor replaced John for three weeks until he was able to preach on April 7. Once well enough to preach, John appears to have continued his busy schedule of preaching in addition to speaking at conferences, conducting funerals, and performing weddings. Mamie must have heaved a sigh of relief when her husband's injuries healed.

A biographical sketch written in 1901 summarized John's previous professional years, and then stated he arrived in Culpeper, Virginia as a fit man for the special purpose of guiding the erection of a handsome church building during his pastorate. John's 1901 biographical sketch included recognition of Mamie's "rare attainments as a charming coadjutor of a minister."

Until the summer of 1901, John and Mamie seemed to vacation together. On July 30, 1901, a Richmond newspaper item mentioned John was be the guest of another pastor at a Newport News cottage. Two days later, a follow-on note mentioned he was spending some time at Virginia Beach. He didn't stay long, however, because by August 6, he was back in Richmond performing a wedding. There is no rest for the weary.

The week of April 19, 1902, Mamie traveled with her father. She and Tom spent a week visiting family friends in Chester, Virginia. I assume Mamie had a household that could run smoothly in her absence. Since I know she had housekeeping help, perhaps John could manage the boys when she was away. During that week, in addition to his already packed schedule, John attended a meeting of his Randolph-Macon College alumni. Perhaps the boys, then ages ten and twelve, were trusted to take care of their own basic needs.

Mamie continued traveling with her father during April 1902. The newspaper account stated they were joined by her father's brother (a medical doctor and long-time Virginia senator), and the three visited what the newspaper called, "the new colored asylum at Petersburg." The paper also printed quotes from Mamie's father who was pleased with the new asylum and its management. Decades after the American Civil War, Mamie and her family continued to make their *all people equal* viewpoints known in word and deed. Absent Mamie's written words on the subject, by accompanying her father and uncle she indicates her agreement with well-documented family sentiments.

A small newspaper article appearing in 1902 gave me a new view of John's ministry. John pastored large, affluent churches yet comments about his sermons spoke less of eloquence and more of man who was warm and caring. When I read he had conducted a funeral at the Home for Incurables where boys from the Orphan Asylum provided music, it was clear John's listeners were not limited to wealthy, influential society members.

Humans (even church people) disagree. According to a May 13, 1902, newspaper article, the organist at John's church resigned because of a misunderstanding with the Sunday school superintendent. John, as pastor, probably listened to both sides then helped restore peace. The last line of the newspaper paragraph read, "Conferences are being held and a friendly adjustment will be made." I wonder if Mamie listened to a replay of those conferences over dinner, then voiced her opinion or offered advice. When John and Mamie visited Richmond eight years later, they were guests of the Sunday school superintendent's widow. Perhaps close, lifelong friendships were God's gift of peace after conflict.

As in other years, John took a summer break. A newspaper announcement indicated he was returning to his pulpit in Richmond on August 31, 1902. At the end of 1902, reports began to appear that after two years in one place, John faced reassignment to another church. Apparently, the people of Park Place Methodist Church campaigned to keep John and Mamie with them until November 1903. Before they left Richmond, however, there were problems in their church.

Under the headline, "No Religion in Richmond," a lengthy quote from one of John's sermons appeared in a January 1903 Richmond paper. John said Christians were becoming inactive and selfish. He said Christianity was lying dead and Richmond needed a resurrection. John said he wanted to see religion rise and rule again in Richmond. Accustomed to preaching to packed houses, John preached his scorching *no religion* sermon to a few attendees and many empty pews. I could almost see smoke rising from the newspaper account of his sermon, so I imagine John's voice was firm when he spoke his thoughts. Was Mamie was thinking, "Amen," or was she squirming in her seat?

What had happened in two years? John hadn't left Park Place in 1902 when his bishop extended his stay; but by mid-1903, he welcomed a change of venue. In August 1903, newspaper reports stated John would leave Park Place for another church assignment. Before the end of November, they were packed and heading out of Richmond.

For the next four years, John pastored Ghent Methodist Church on Raleigh Avenue in Norfolk, Virginia. They arrived as the church was enjoying a new building. I visited beautiful Ghent in 2008 and found it still vitally alive. Patricia, an administrative secretary, graciously located pertinent records of the years John pastored there, and corresponded with me to provide an update after my visit.

The secretary of the Sunday school warmly recorded John's arrival at Ghent. "We have known our new pastor from his boyhood," the Sunday school secretary wrote. "The purity of his character, the sublimity of his faith, and his relentlessness against sin make his presence among us valuable to the church and the Sunday school. The man in all relationships is most lovable, kind and sympathetic. Wherever he goes, his influence is for the best." If Mamie's husband exhibited an equally caring nature in private, he must have been a loving husband.

Mamie and the boys joined Ghent Methodist Church on November 29, 1903. John's brother, William, joined Ghent on May 22, 1904, followed by John's mother on June 19, 1904. William had finished his job assignment in New York, and had moved back to Norfolk to live with his mother. Family members joined churches John pastored if they lived close enough to do so. Since Mamie seems to have had good relationships with John's family members, their support of John's ministry must have pleased her as well.

On October 18, 1904, Mamie spoke at the Foreign Missionary Society meeting in Danville, Virginia. Apparently, some women addressed the gathering, followed by other women giving responsive speeches. Mamie, identified as the wife of a former Danville pastor, spoke in response. The three-day session was a gathering of one hundred delegates from all over Virginia. Mamie enjoyed time with other pastors' wives and visiting women she had known during their earlier Danville pastorate.

During his pastorate at Ghent in Norfolk, John performed eleven marriages. One wedding joined his niece, Roschelle, and her fiancé, Theodore, on November 30, 1905. On April 9, 1907, he baptized their daughter, Irene. He performed ten additional baptisms at Ghent. The church received 265 new church members during John's four-year pastorate. Mamie must have been busy trying to meet and help welcome all those new members.

Cousin Patsy gave me two copies of a photo postcard that provided insight into the daily life Mamie and her sons were living during their time in Norfolk. Half of one card survives with the photograph on one side, and the reverse bearing only the address of Mamie's younger sister, Sallie, at a Conservatory of Music in Durham, North Carolina, where Sallie was studying or teaching. The other uncut copy includes Mamie's message and the address of her father and this third wife.

Taken in late 1906, the photograph captured Mamie's sons, Robert (seventeen) and George (fifteen), seated on the front steps of a house with a beautiful Collie. Both boys are dressed in what appear to be school uniforms with ties and caps. Mamie's note reads, "Isn't this good of the boys? They were spending the day with one of their schoolmates and he took their picture. The dog happened to be passing by and they decided to have him in the picture too. George is scratching his (the dog's) head to keep him still. The sun was shining in their eyes and caused them to frown. Our love to all, Mamie. Jan. 15, 1907."

Based on what I know about Mamie's fastidious housekeeping, I doubt her boys ever had dogs of their own, so I'm amazed to have a photograph of my grandfather with a dog. After reading Mamie's words, I looked closely; the tips of George's fingers are indeed visible on top of the pretty Collie's head. Mamie apparently thought a photograph of her sons pictured with a pretty dog was something her family members would enjoy, so she made the effort to get copies into their hands. I'm sure their aunt and grandfather were pleased to receive the pictures as well as the description of how the moment captured in the photo came about.

John and Mamie's association with Ghent Methodist Church didn't end permanently when they moved to their next charge. Later, Ghent was among the group of churches John served as Presiding Elder from November 1931 until his death in 1936. After visiting Ghent myself, I smiled to think how pleased John and Mamie would be to see that church still flourishing so many decades later.

John returned to Richmond for such events as weddings. Perhaps girls who grew up during John's pastorate wanted him to perform their weddings. Mamie's name often appears among the names of wedding guests, so she must have enjoyed those trips back to former locations to renew friendships as well. Although based in Norfolk, John continued his practice of conducting revivals and camp meetings at other churches around Virginia.

In October 1907, after John had spent four years at Ghent, it was time to move again and John was awaiting word about their next assignment. In November 1907, the Ghent records indicate the removal of Mamie,

George and Robert from membership; they followed their husband and father to Market Street Methodist Church in Petersburg, Virginia. They were moving about eighty miles back into the interior of Virginia. They were also moving away from John's Norfolk-based family and farther from Mamie's Eastern Shore family.

If they were sad to leave Norfolk, Mamie probably appreciated the warm welcome they received in Petersburg. Officers of Market Street Methodist Church met them at the railroad station and took them to the parsonage on the evening of November 30, 1907. After the holidays, on January 3, 1908, the church membership welcomed John and Mamie with an informal reception hosted by the women's missionary society of the church. The missionary society president assisted John and Mamie in the receiving line. Imagine Mamie and John struggling to remember all the names. The missionary society members decorated the church lecture hall for the reception, and arranged a program of vocal and instrumental music. How kind of their new church to welcome them, and make the beginning of their Petersburg assignment pleasant.

During John's Petersburg pastorate, he continued to speak in venues outside his own pulpit. During his first weeks there, he spoke to the Methodist Sunday School Union of their area. John conducted revival services for a Rev. Winn at Boydton, Virginia in May 1908. Rev. Winn was married to Mamie's first cousin, Zillah. Zillah's mother, Maggie, was the sister-in-law recipient of many letters written by Mamie's mother and included in Cordelia's chapter. If Mamie and Zillah were as close as their mothers had been, Mamie was delighted to accompany her husband on that revival trip. If I hadn't researched Mamie's cousins and their spouses, I might have failed to realize Rev. Winn was Mamie's relative by marriage when I found the revival announcement.

In addition to a busy summer that included conferences, John and Mamie managed an August visit to their friends and family in Norfolk and on the Eastern Shore. During their Eastern Shore visit, John preached at the Presbyterian church in Belle Haven. It seemed odd for Presbyterians to invite a Methodist to preach, until I learned Mamie's sister and brother-in-law were members of that church.

They were still in Petersburg when Mamie's father died at his home in Accomack County, Virginia, on March 19, 1909 at the age of seventy-seven.

His passing was a great emotional loss to his children and grandchildren. Tom's gravestone in the little family cemetery near Atlantic View reads, "In Memory of our Honored Father, Asleep in Jesus." Tom was never idle until age and illness slowed him in the last years of his life. After he retired from his produce business to live on the Eastern Shore in the winter of 1895, he kept busy. After his death, several individuals wrote stories of his accomplishments.

Throughout 1910 in Petersburg, John continued his busy schedule with many services, weddings, and funerals. He delivered the invocation and benediction at a special Memorial Day service with the Daughters of the Confederacy in attendance. Between John's opening and closing prayers, the Petersburg Grays fired volleys in honor of veterans. After John spoke on *The Spirituality of the Church* at a district Methodist Conference in July, he and Mamie again took their family on a seaside vacation.

Mamie and John hosted a wedding in their parsonage in September 1910. Later that month, John assisted another pastor in performing the wedding of Mamie's niece, Eva, in Onley, Virginia, on the Eastern Shore. Eva was a daughter of Mamie's sister, Bessie, who had died when Eva was eleven years old. Mamie couldn't have imagined only fifteen years later, she would attend Eva's funeral.

John must have been working too hard. On September 27, 1910, a newspaper mentioned he had returned to his duties after being in delicate health for several weeks. Even illness didn't appear to slow him down. In November, John wrote reports on tasks he'd performed for his Methodist conference. John also continued appearing as a guest pastor in other churches. John's hectic schedule and precarious health must have worried Mamie.

In June 1912, Mamie's younger son, George, received his undergraduate degree from Randolph-Macon College in preparation for Law School at the University of Virginia. John had a reason to return to Randolph-Macon that June as well. Randolph-Macon College conferred upon John the Doctor of Divinity degree in recognition of his service to the Methodist church. On June 26, 1912, George asked a pretty girl named Marie to become his wife after he finished law school. In August 1912, John's mother passed away. It was a busy summer.

Mamie's older son, Robert, lived long enough for me to have known him, and I regret I didn't meet him. A mystery I wish I had explored with Robert (aside from the disappearance of his brother) was Robert's failure to attend the Naval Academy. September 1908 newspapers mentioned Robert's nomination by a Congressman to attend the United States Naval Academy at Annapolis, Maryland. Later, another newspaper article stated he had departed for the Academy although there is no record he ever attended. Perhaps a health problem prevented his military service. He attended college somewhere, however, because the 1910 census lists Robert as a twenty-year-old student. By 1915, he was working as a clerk for the railroad. He continued with the railroad as an auditor, investigator, and claims agent. Robert and George also owned an automobile company for a short time.

George's Randolph-Macon college years followed by law school at the University of Virginia at Charlottesville were marked with success. He earned orator's medals on debating teams at college and university. He was a member of Sigma Phi Epsilon fraternity as well as literary and debating societies. I'm sure Mamie anticipated successful lives for both sons. She was probably beginning to feel she'd completed her tasks as a mother and was adjusting to the emptying of her nest.

The 1913 Norfolk/Portsmouth Directory lists John as pastor of Monumental Methodist Church, and the family was living on Court Street. During that pastorate, Mamie's brother, George, passed away on September 4, 1913, in Erie, Pennsylvania. With his passing, Mamie became the eldest living child in her family. Thankfully, the grief of sibling deaths was over for the rest of Mamie's lifetime. Susie, Eva, Emma, and her half-sister, Sallie, outlived Mamie.

On November 11, 1915, Mamie's younger son, George, married Marie, whose chapter you've read. Marie's father was also a Methodist minister and the wedding took place in her family's Methodist parsonage in Norfolk, Virginia. The fathers were well-acquainted colleagues in Methodist ministry. Both families surely believed the pair well matched. Unfortunately, I never asked Marie about her relationship with Mamie, so I don't know how well the two women got along. During the early months of their marriage, George and Marie were living within a half-mile of John and Mamie's home and probably attended John's church. Mamie had about three months to become acquainted with Marie.

By February 1916, John was serving Main Street Methodist Church in Suffolk, Virginia. Correspondence in the 1914-1916 timeframe indicates the church was actively involved in supporting prohibition. As good Methodists, pastors encouraged their flocks to vote *dry*. I noted that John, when writing to his fellow Methodist leaders during this period, often included his wife's "kind regards." Those years were busy times for John and Mamie because their congregation was building a new facility and experiencing growth.

When John asked his bishop to preach at a special service in February 1916, John invited him to stay overnight in John and Mamie's home. John's use of the word "again" in his letter indicates the bishop stayed in their home on previous visits. Having guests meant extra work for Mamie even if she had help with her home. At least fellow ministers understood the lack of luxurious accommodations.

In 1919, George must have fulfilled a vision Mamie had for her younger son when she named him. Her uncle George had served Virginia as a senator and her son, George, was to serve in the Virginia House of Delegates representing Portsmouth, Virginia. A lawyer, George was working as a city attorney in Portsmouth at the time of his election.

On March 17, 1920, Mamie and John's only grandchild was born. John's status as his grandfather's namesake should have placed the little guy in good stead with his paternal grandparents. George continued his law career and served the first of his two sessions in the Virginia House of Delegates. Mamie must have been proud of her son.

The early 1920s should have been pleasant years for Mamie and perhaps they began that way. John was serving as pastor of the beautiful Larchmont Methodist Church in Norfolk, and their home was nearby on Cambridge Crescent. John's brother, Robert Isaac, lived down the street with his wife and large family.

On July 4, 1921, Mamie and John's younger son and his family came to visit for the day. Leaving Marie and baby John with his parents, George announced he was going to walk to the store for cigars. He left and never returned. He failed to serve the second of his two sessions in the House of Delegates. He abandoned his wife and child. After several years of my own detective work, George's story is still a mystery. John and Mamie must have found such an odd occurrence to be not only crushing because they had lost their son, but also embarrassing as parents and people in spiritual leadership. Newspaper accounts reported the issuance of a lunacy warrant. They must have thought George had lost his mind.

In 1925, Mamie's older son, Robert, married a woman named Roselyn who also worked for the railroad. Robert and Roselyn lived in the neighborhood where his parents and other relatives had made their homes. They did not have children.

In the 1926-1927 timeframe, John and Mamie were again on the Eastern Shore at Salisbury, Maryland. John was serving as Presiding Elder over a group of churches rather than pastoring an individual church. Perhaps resulting from the strain of their son's disappearance, John requested a less demanding assignment without full-time pastoring of a congregation. For Mamie, living near her siblings and their families provided a healing balm.

In 1930, John was again pastor of Larchmont Methodist Church. The assignment meant Mamie and John could move back to their previous home on Cambridge Crescent in Norfolk. Living in that neighborhood wasn't the same for John because his oldest brother, Robert Isaac, had passed away. The family lost another member when Minnie, the wife of John's youngest brother, William, died suddenly on September 28, 1933 at the age of fifty-two. Since she and William were members of their close-knit family as well as members of the church John was pastoring, it was a sad time for everyone on all fronts.

John served at Larchmont until he retired from active ministry in 1933. He remained involved in Methodist causes, and they lived in the neighborhood adjacent to the church until John's death in July 1936. John's death wasn't a sudden shock for Mamie. He was ill for three months and hospitalized for three weeks prior to his death. After forty-seven years of marriage, Mamie faced life without her husband. After John died, a glowing report of his fifty years of ministry called Mamie a godly woman who had been an inspiration to her husband.

John's obituary listing of family members who survived him included not only his son Robert but also his son George who had been missing for fifteen years. Since I feel certain Mamie wrote or at least agreed with the wording of John's obituary, I wonder if she continued to hope George was alive, even though his whereabouts were unknown.

During her widowhood, Mamie continued her summer trips to the Eastern Shore to visit relatives and vacation at the Wachapreague Hotel.

Cousin Emily said the family referred to her as *The Grand Dame of the Wachapreague Hotel* when she came to visit. Mamie had Parkinson's disease, and eventually required a nurse to accompany her to provide the necessary assistance Mamie's deteriorating body required.

Seeking a better understanding of Mamie's illness, I learned Parkinson's disease is second only to Alzheimer's among neurodegenerative diseases. Four major characteristics of the disease include tremor, slow movement, stiffness, and poor balance. Apparently, most victims of the disease notice tremor in a limb first, with symptoms beginning on one side of the body, and then spreading to include all parts of the body. The disease produces stooped posture as it progresses.

Small, illegible handwriting can be another symptom of Parkinson's disease. Mamie's 1907 postcards confirm that symptom. In 1907, her handwriting was still legible, although it was quite small. Mamie's voice could have taken on an unpleasant monotone quality that caused her to curtail her speech. Progressively worsening symptoms typically cause anxiety and depression.

Unlike her surviving sisters who retired to the Eastern Shore to live out their lives as close neighbors, Mamie continued to live in Norfolk, and visited the Eastern Shore during the summers. Mamie's unhappiness in her later years possibly stemmed from dissatisfaction with her life. She could have thought her sisters made better life choices.

When I researched the lives of Mamie's sisters, I wondered if Mamie wished she'd been a career woman. One full sister, Susie, had a long, eventful career in education. She taught high school and college then served as a high school principal. A few years after the nation's women gained the right to vote, family correspondence mentions Susie's selection as one of five women to serve as the first female delegates to the Virginia state democratic convention. Mamie's half sister, Sallie, had a career in music, and did not marry. There must have been days, especially after George disappeared, when Mamie felt she failed in mothering her younger son. On those days, Mamie could have envied the two sisters who chose to live their lives as single, independent women.

Another full sister, Eva, married a dentist, had no children, and lived in Belle Haven, Virginia, where she apparently enjoyed life near her siblings and extended family. In Susie's later years, she built a little cottage behind Eva's home where she lived until her death a few months after Mamie died. Mamie's youngest full sister, Emma, was married in 1907 in a ceremony performed by John in a church near the family ancestral home. Emma purchased Atlantic View from her siblings, and then raised her family there. Did Mamie wish her husband's career permitted them to live on family land and stay in one place? Perhaps.

After Robert Isaac passed away in 1926, Mamie's sister-in-law, Annie Irene, continued to live in the large home where she and her husband had raised seven children. Perhaps because she had the space, she invited several family members to share her home in their declining years. According to Annie Irene's descendants, a staff of maids, a butler, and a chauffeur supported the household. Living in one household allowed family members to pool their funds and live better as a group. Mamie lived part of her widowhood with Annie Irene until her health deteriorated and she entered a nursing home. Even Mamie's sister, Susie, lived in Annie Irene's household during her breaks from teaching.

External forces also played a role in the merging of family members into one household. The influx of over one hundred and fifty thousand military personnel staffing the naval activities in the Norfolk area created housing shortages in the early 1940s. Whatever financial and housing needs made living together necessary, I'm sure the emotional security of living with much-loved relatives was an added benefit as Mamie's generation lost their spouses. Annie Irene was a wonderful sister-in-law.

Why didn't Mamie live with her son, Robert, after John's death in 1936? The 1930 census indicates Robert's household included his wife, Roselyn, and Roselyn's sixty-year-old mother. Perhaps because Robert already housed his wife's mother, Mamie didn't want to burden her son and daughter-in-law.

Reflecting again on Mamie's lack of involvement in her only grandson's life, perhaps by the time her husband retired and she might have had more time, she could have been too ill to help care for her grandson. The situation is further complicated by distance; Mamie was in Norfolk and her grandson was in Richmond.

Mamie's health eventually deteriorated and nursing home care wasn't enough. When she passed away on July 14, 1944 at the age of seventy-five, she was in a Norfolk hospital. Her obituary stated she was a native New Yorker and widow of a well-known Methodist minister. She was a member of Larchmont Methodist Church (the last church John pastored) at the time of her death. Her obituary stated her son, Robert, and grandson, John (my father), survived her along with four of her sisters. If Mamie continued hoping her son, George, was still living when his father died in 1936, her son Robert did not believe he was alive when she died in 1944. There was no mention of George at all.

Mamie's obituary noted her grandson, John, was living in Greensboro, North Carolina. When he and my mother married in 1942, I don't believe there was any communication to or from Mamie. The absence of a grandmother's communication upon the marriage of her only grandson is inexplicable. Mamie's obituary proves she knew where her grandchild was living.

Based on my father's lack of interest, I'm guessing his mother, Marie, had kept Mamie and Robert informed, even if they never asked about John. Marie must have dutifully continued to write polite letters to her mother-in-law and brother-in-law. Even if Mamie's handwriting and speech had deteriorated from Parkinson's by the time of my parents' 1942 marriage, my father's uncle, Robert, could have communicated on her behalf. There was apparently no contact in 1942 and I know my parents did not attend Mamie's funeral in 1944. How odd.

After Mamie's funeral service and burial next to John at Forest Lawn, her son and daughter-in-law, Robert and Roselyn, hosted a luncheon for the family at their home. Mamie's family from the Eastern Shore traveled to Norfolk for her funeral and the family luncheon. Mamie would have loved that gathering. Mamie's half sister, Sallie, made note of Mamie's death in her diary. After stating the date of her death and funeral, Sallie wrote, "Sister Mamie looked very peaceful after her long illness."

When Cousin Patsy was showing me that page, she pointed to a note later added by Sallie. As if it were a connected follow-on to Mamie's death, the note mentioned the birth of Mamie's great-grandchild in 1947. I am that child. I experienced a surprising rush of emotion when I saw those words. Seeing our names on the same page made me feel connected to Mamie. Sallie, by recording my birth on the page where she had recorded Mamie's death, seemed to wish Mamie had lived to know she had a great-granddaughter. Mamie died almost three years before I was born. I'm sorry we didn't cross paths.

At Mamie's death, Robert inherited John and Mamie's estate. Since I did not meet Robert and he had no children, I've been unable to learn the disposition of items I would have treasured such as John's sermon notes and Mamie's personal items such as her Bible.

Having rushed through the many moves and personal events of Mamie's life, let's slow our pace to explore public commentary about John and Mamie. If John's home behavior mirrored his public persona, Mamie had chosen her husband well. Extended family members, congregations, and fellow ministers repeatedly used words such as, "grace, kindness of manner, sensitive human understanding, and healing" when they spoke of John's dealings inside the church, and his treatment of other people in the communities where he served.

After John's death, the *1936 Journal of the Virginia Annual Conference of the Methodist Church* included poignant statements about the love his fellow ministers felt for him. They reportedly missed his "beaming face, which was the outward symbol of the gracious spirit which tenanted his body." They wrote of his love for others, and referred to John's efforts to reveal the heart of God to people who didn't know Him as Savior. Descriptions of John mention his beautiful, kindly, Christ-like spirit. John's fellow ministers, as part of their tributes to John, describe Mamie as an attribute. She was, they wrote, a godly woman who became the inspiration of his life and the mother of his two children.

Did John's professional friends mean what they said about Mamie, or were they being polite? Family memories of her aren't as glowing. Mamie's husband and sons feared her stern reproach. Cousin Emily, who said she was afraid of her aunt Mamie when she was a child, said Mamie demanded her husband and sons remove their shoes before entering their home.

By stark contrast, Mamie's family members who knew John are quick to speak of his warm, loving personality. I assume John's public reputation as a man of kindness must have been accurate and in consonance with his private persona. All reports about John from Mamie's family have been consistently positive. "We loved Brother John," Cousin Emily told me.

My own assessment, however, raises a question. Why, if John were such a loving person, did he not make an effort to have a relationship with his grandson? From the few times they together, my father remembered

John as a kind, smiling man. John lived for sixteen years after my father's birth and could have stepped in as a father figure had he chosen to do so. John's obituary stated he was only ill for three months prior to his death at seventy-three, so he must have been healthy enough to spend time with his grandson. He was retired for the last few years of his life. Why did he miss the chance to be a positive influence in his grandson's life? Even if Mamie was unwilling, physically unable, or too hurt over the loss of her son, why didn't John do something?

Perhaps mental aspects of Mamie's Parkinson's symptoms came into play. She could have experienced depression or dementia. Did she pass along a mental issue to her son, George? Whatever happened, her husband apparently stood by her. If John endured difficulty at home, God used the situation to prompt the sympathy and kindness John exhibited to others.

What can I learn from Mamie? My own need for solitude and peace has left me unable to graciously cope with other people at various times in my life. I've spoken too quickly and without kindness. I've been impatient. When overwhelmed by too many people demanding too much, I withdraw. Perhaps you feel guilty of the same failings. It is never too late to change if we are physically and mentally able.

If we're ill, let's seek help. In addition to medical treatment or counseling, we all benefit by seeking God's help. Scripture suggests we turn impossible situations over to God. By faithful study of God's principles, we are better equipped to rise above what is bothering us and live more positively.

During a lifetime of excellent, Bible-based teaching, could spiritual transformation elude Mamie? Yes. Merely sitting in church did not give Mamie the power to triumph in life, or the privilege of spending eternity in the presence of God. Did she fail to tap into the power of God? Did she live without the one true, lasting, source of wisdom and strength by failing to accept Jesus as her Savior? Did she give lip-service acceptance of Jesus; then deny the Holy Spirit residence in her soul, and attempt to live by her own power? Let's be certain of our salvation, and then live empowered by the Holy Spirit.

I trust Mamie did receive salvation and she's spending eternity in God's light. As much as she loved her ancestral home, Atlantic View, her new home has an even better view. Joy has replaced Mamie's physical, emotional, and spiritual struggles and she finally knows what happened to her son, George. She lives where loved ones never die or leave; they arrive and stay forever. After a lifetime of packing and moving, Mamie is finally, eternally, at home.

Daisy

1873 – 1965

Grandma Daisy, my great-grandmother, lived until I was seventeen years old so I knew her well. I visited Daisy's home, ate at her table, and cherished every moment with her. Daisy lived in Virginia while I lived in New York, Colorado, and North Carolina, so our visits were infrequent yet memorable.

A 1950 four-generation photograph taken outside Grandma Daisy's home in Richmond, Virginia, documents our first meeting. Daisy was seventy-six and I was two years old when we were photographed standing with her daughter, Marie (my grandmother), and Marie's daughter-in-law, Grace (my mother). Someone also snapped an informal photo of Daisy and me. Priceless. During a 1953 visit with my grandmother in Richmond, we again spent time with Daisy.

I last visited Daisy during the 1959 Christmas season. My mother and I traveled by train from North Carolina to Richmond, met my grandmother (who then lived in Maryland), and we all stayed with Daisy. Then twelve years old, I was interested in baking (probably because the process included consuming the results) and Daisy's baking made a lasting impression.

More than fifty years later, I clearly remember Grandma Daisy standing in her tidy little butler's pantry, opening a tin of cookies, and offering the contents to me. The tin contained perfect little Christmas sugar cookies. Selecting a little tree, I bit into the buttery cookie coated with an ultra thin layer of decorative glaze. I looked up into her sweet face and smiled with pleasure. We spoke volumes but exchanged no words. How perfect was that? I have made thousands of sugar cookies since that day and Grandma Daisy's cookies remain the ultimate standard of excellence.

Between visits, Daisy was a faithful correspondent. Her children, grandchildren, and great-grandchildren were fortunate recipients of her letters. She faithfully sent gifts and cards (bearing a little money) on our birthdays and at Christmas. In one belated birthday note to me, Daisy apologized that my card and gift were late. She explained she had been ill, and while she recuperated in her second-floor bedroom, someone broke into her home and stole some sterling silver flatware. Typical of her selfless nature, Daisy glossed over the burglary story, but expressed great concern because she'd missed my birthday.

Known as *Daisy*, *Mama*, and *Grandma Daisy* to various generations, her active love and sweet disposition attracted family and friends. She was petite in form and strong in spirit. Daisy was much more than a great cookie baker as you'll learn in the pages of her story.

Daisy was born on January 9, 1873, in a tiny hamlet originally called Tobacco Stick in Dorchester County on the Eastern Shore of Maryland. Creeks and inlets off the Chesapeake Bay surround Daisy's childhood community, later called Madison. Daisy was the firstborn of two widely spaced children.

At birth, Daisy may have had no grandparents to welcome her. Her mother's parents had passed away. Daisy's paternal grandfather, Zachariah, disappeared from census records after 1840, and was reportedly lost at sea. Zachariah's wife, Leah, disappears from census records after 1870; but may have been living at the time of Daisy's 1873 birth.

Daisy's father, Luther (*Captain Luke* to his neighbors), began his working life as a sailor. In his area, men who once captained a boat or ship gained the lifelong title of *Captain*. Luther's job description as a sailor first appeared in the 1870 census when he was single and living with his mother. When Daisy was eight years old, the 1880 census noted Luther was again working as a sailor. Luther eventually owned two schooners. By 1900, ten years after Daisy had married and left home, the census records Luther's occupation as merchant. According to family lore, he became a merchant shortly after 1880. He continued to sail his schooners to Norfolk and Baltimore to deliver lumber from the Eastern Shore, then return with merchandise for his general store. Luther provided well for his family.

Viewing Daisy's busy ninety-two years in three major phases of childhood, marriage, and widowhood, the shortest segment is her childhood prior to her marriage at seventeen. Daisy's chapter will quickly mention her childhood then focus primarily on the details of her forty-seven years of marriage, followed by her widowhood of twenty-seven years. Further details of Luther's business life, Daisy's ancestors, and her early environment are included in the story of her mother, Sarah.

Since Daisy lived a long, healthy life, diet during her formative years merits mention. People in Daisy's home county had a wonderfully varied diet. Apricots, peaches, pears, persimmons, blackberries, walnuts, and hickory nuts grew nearby. Her family grew or had access to local crops such as sweet potatoes, corn, wheat and oats. Women served seafood, chicken, beef, and pork with cornbread and biscuits. Wild ducks and geese were abundant and frequently served as part of local meals. Daisy, as the daughter of a merchant, enjoyed items grown elsewhere (such as coffee, tea, and sugar) that her father purchased on his trips to Baltimore and other ports.

Families in Daisy's hometown placed great importance on education. They built a schoolhouse and hired a schoolmaster, so Daisy and other children could attend grammar school. When Daisy was about fifteen, she continued her higher education at a female academy in Baltimore. Her school was too far from home to commute daily, so she boarded in a hotel. Her father whisked her home when she mentioned falling in love with the much-older hotel owner. Her parents were seeing the first signs of Daisy's destiny. They stopped her first romance, but Daisy would marry an older man.

Daisy's brother, Oden, was born on December 22, 1889, eighteen days before Daisy's wedding. Although they didn't grow up together, Daisy and Oden had a loving relationship and were close throughout their adult lives.

At sixteen, Daisy was engaged to Vincent, a Virginia-born Methodist minister. Vincent served a circuit (group) of churches including the little church in Daisy's hometown. He had begun his ministry in 1884 in the East King and Queen Circuit in Virginia. In 1886, he arrived on the Eastern Shore of Maryland to serve churches in Dorchester County, where Daisy lived. Something about that preacher attracted pretty, petite, Daisy.

Perhaps Daisy first noticed Vincent's physical appearance. A description of Vincent in Methodist documents about the time of their marriage stated, "He has many advantages in his physical makeup. He is in weight and height of good proportions, firm in limb, erect, and of exceptional completeness in form and features. His hair is jet, eyes blue; he wears a moustache of light-brown color. His well-chiseled face invites confidence. He is a young man of unusual promise, modest, ready for every good word and work, and ambitious to be accepted of his Lord."

Vincent's charismatic sermons might have won Daisy's heart. Vincent had been in the pulpit six years, so he had time to develop his preaching style and feel comfortable delivering God's message. Later descriptions of his sermons mention three recurrent themes: the dangers of sin, man's personal responsibility to God, and the glory of life in Christ. He loved expressing the truth, "the gift of God is eternal life through Christ Jesus." No audio recordings of Vincent's sermons exist, so I've not heard the voice Daisy heard. According to a fellow minister, Vincent "brought individuals and families by the hundreds to accept the Savior, and follow the way He laid down for us." From that description, I imagine Vincent's speech pattern was pleasing to the ear and his manner gently persuasive.

A man of good manners, I assume Vincent asked Luther for Daisy's hand in marriage. Imagine Luther (then forty-three) contemplating the marriage of his young daughter to a man only ten years his junior. Luther encouraged the match. Vincent's personal character and faith, his ministerial career, and the good reputation of his Richmond family must have earned Luther's approval. Daisy celebrated her seventeenth birthday by marrying Vincent, then thirty-three.

Aside from good looks and a pleasing preaching style, Vincent was well educated. He attended private schools in Richmond followed by his studies for the ministry at Randolph-Macon College. Vincent was not, however, wealthy. Before Vincent began his career as a minister, his family lost much of their once-plentiful, hard-earned money during the American Civil War. When he met Daisy, Vincent had a horse, a buggy, a suitcase containing a limited wardrobe, and an annual income of $305. Clearly, she did not marry him for his money.

The newspaper wedding announcement titled, "Marriage of a Preacher," identified Daisy's parents and mentioned Vincent's ministerial position. They were married in St. Paul's Methodist Church in Cambridge, Maryland, at seven-thirty in the evening on Thursday, January 9, 1890. The ushers, bridesmaids, and attendees were Daisy's friends and family. No members of Vincent's Richmond family attended. Travel across the Chesapeake Bay to Cambridge on Maryland's Eastern Shore was probably too difficult and costly for his parents.

The announcement included a description of Daisy's wedding attire that read, "The bride wore a dress of white satin en train with short sleeves, and the tulle veil caught up in a wreath of orange flowers. She carried a bouquet of La France roses." Following the ceremony, the bridal party and relatives of the bride were entertained at an informal reception at a friend's home near the church. The article stated the newlyweds, after a northern tour (which I hope included Richmond to visit Vincent's parents), would reside in Cape Charles City, Virginia. Vincent was to preach in a church in Eastville, Virginia, and serve as the first minister for a new mission at Cape Charles.

The slow financial recovery of the family business in Richmond could have enabled Vincent's father to provide some funding for the newly wed couple. Vincent's mother, Therese, was a giving person who probably provided whatever household goods she could to help the young couple furnish their nest. Daisy's merchant father provided goods from his store when the couple visited or lived nearby. Daisy appreciated any supplement to Vincent's meager salary.

Daisy was a frugal housewife who found ways to stretch their money. Children and grandchildren who grew up in Vincent and Daisy's home said when they asked Vincent for money, he often emptied his pockets into their hands. They did not ask Daisy, who carefully scrimped and saved for necessities and was seldom able to hand over cash for frivolous extras.

Stepping into the role of minister's wife at the age of seventeen could not have been easy. Older women in the congregations Vincent served during those early years might have viewed Daisy as a child. We want to believe church members are too spiritual to behave unbecomingly; realistically, we know Daisy experienced difficult moments during her long career as a minister's wife. To paint a quick picture of Daisy's married life, I'll recount Vincent's moves from church assignment to church assignment; annotated with the arrival of children and other major milestones. After we've followed the trail of her husband's career, we'll revisit Daisy's role. In general, at every church assignment, Daisy was supporting Vincent's work, listening to his problems, and raising their family.

For almost two years, Daisy and Vincent enjoyed married life without children on the scene. Then on November 15, 1891, Daisy gave birth to her first child. Little Vincent (whose middle name was Daisy's maiden surname) was born when they were still assigned to the Eastville/Cape Charles ministry, and his birth is recorded in Accomack County, Virginia. I'll refer to their son as Vincent Jr. even though he had a different middle name. Daisy's responsibilities included keeping house, raising her baby, and being a minister's wife with all the associated social and spiritual duties.

In 1892, Vincent's assignment returned them to the Dorchester Circuit near Daisy's parents. When my grandmother, Marie (named for Vincent's mother), was born in December 1893 her birth was recorded in Madison, Dorchester County, Maryland. Vincent's 1893 transfer to Fox Hill, Virginia, occurred too late in Daisy's pregnancy for her to travel or she returned to have her second child with her mother's help. Vincent Jr. was two years old so Daisy had her hands full with a toddler and a newborn.

From Fox Hill, Vincent went to Smithfield, Virginia. Next, they went to Churchland, Virginia where Daisy Emma (called Emma, the middle name of Daisy's mother) was born in August 1896. Daisy must have begun to realize frequent family moves made life challenging. Finding a place to live was apparently not an issue. Their children have said churches provided a house near the church or a parsonage next door. Packing, moving, and setting up housekeeping repeatedly, must have made Daisy want to limit her possessions. Not one to complain, Daisy must have experienced instances when the church-provided housing wasn't right for their family's needs, and she simply had to manage.

In 1900, Vincent's next assignment was Capron, Virginia where Daisy's fourth child, Grace, was born in 1901. Vincent moved the family to South Norfolk where their last child, Walter, was born in 1906. In 1907, they moved to pastor Liberty Street Methodist Church in Portsmouth, Virginia.

In 1908, Vincent was pastoring Denny Street Methodist Church in Richmond. Vincent's mother, Therese, died in Richmond in August 1908. I hope Vincent and Daisy arrived in Richmond early enough to spend time with Therese. It would have been gracious of Methodist leaders to relocate Vincent near his mother, if they were aware she was ill. Perhaps Vincent's presence was a comfort to his father, who died in May 1909. Living in Richmond afforded Daisy the opportunity to know her husband's extended family members, whom she grew to love. Perhaps Daisy enjoyed Richmond life because she and Vincent eventually retired to Richmond.

Vincent's Richmond assignment was, according to written accounts, the place Vincent accomplished his greatest work. His skill as a church starter influenced his assignments. Vincent could explain the plan of salvation in a manner that encouraged people toward conversion, and then he could organize churches to support the spiritual growth of new converts. At Denny Street Methodist Church, Vincent realized many poor people living near their church needed more than spiritual food. They needed physical food and clothing, medical attention, and jobs. Absent secular social systems, churches were attempting to fill those needs. Vincent helped organize classes in nursing, homemaking, and sewing. If not directly involved in teaching, Daisy surely helped Vincent identify skills most useful to the women.

Vincent organized and taught a Bible study class of one hundred men as part of the social outreach effort. The church auditorium became too small to accommodate the growing congregation, so Vincent spearheaded plans to build a larger building. After he planned new buildings and raised funds, the Methodists reassigned Vincent to create or energize another congregation, while his successors enjoyed the buildings he had envisioned.

Vincent's next challenge was to convince leading men at Denny Street Methodist Church as well as members of surrounding Richmond churches, to help erect a church in Norfolk, Virginia, his next assignment. His successor completed the new Norfolk auditorium Vincent had envisioned. Before he left Richmond, he also had a vision of erecting a Union Chapel in the Fulton Hill section of Richmond. He interested leaders of several denominations in the joint effort to serve that rapidly growing section of Richmond. The vision became reality; he soon dedicated a chapel where they held Sunday school weekly, and where ministers of several denominations alternated preaching duties on Sunday afternoons. Many years later, a pastor writing about Union Chapel, stated that all churches later founded in the Fulton Hill section came into existence as an outgrowth of Union Chapel and Sunday school.

Before leaving Vincent's professional life, let's explore the character depths of the man Daisy married by reading an assessment of Vincent written by a leader of his denomination: "The outstanding characteristic of Vincent was his clear conception of the value of personality. He saw in everybody he met, however low, the pearl of great price, the ability to become a child of God. He realized the greatest work any man can do is to take these lost children of God, (and) as did the Savior, seek them out and bring them into the fold. His has such a clear vision of what the kingdom of God could be, that there grew in him a divine discontent; which ever drove him by personal work to endeavor to bring the kingdom of God into the lives of the lost."

According to a tribute written after Vincent's death, he began encouraging people toward a personal relationship with the Lord while in seminary. One evening Vincent asked a friend, "Are You a Christian?" The young man replied that he had wanted someone who knew something about Christianity to ask him that question. After talking past midnight, the two young men parted. At daybreak, the young man was pounding on Vincent's door saying, "I've got it!" The young man later became a judge and a strong church leader in Florida. Vincent never hesitated to speak to adults or children about their soul's welfare. He approached them on the street, entered their homes, and was instrumental in bringing hundreds to accept the Savior, according to his fellow ministers.

Some of Vincent's family members and their friends thought Vincent had given up potential wealth, social position, and a ready-made future in his Richmond family's business. A minister said Vincent gave up all, "to serve the Lord as an obscure Methodist itinerant." The minister went on to say, "The Lord gave all these things back to him and more."

Vincent's father expected his sons to join the family business, as he and his brothers had joined their father's business a generation earlier. Vincent's older brother, Charles, had been the obedient son. Charles must have felt he had no way out. Of four brothers who grew to adulthood in Vincent's family, only Charles spent his career working in the family business. Vincent's brother Robert had died as a young man who had barely begun his working career, his brother Victor left home to live in Idaho, and Vincent chose the ministry (or felt God had chosen that vocation for him).

Charles worked diligently in the family business and asked his father's help to begin his own business venture. In letters to his fiancé, Charles mentioned looking forward to the day when Vincent was old enough to perform his role in the business. No matter how much Charles loved Vincent or thought his calling to the ministry a valid choice, it had to be disappointing for Charles to set aside his own dreams. Would Daisy hear of resentment between Charles and Vincent during the early years of her marriage?

If Charles did feel resentful, his Christian values helped him keep silent. As a teenager, he had worked full-time in the family business while he attended college at night. He was frequently ill because his job entailed walking to visit customers in inclement weather. When it was Vincent's turn to go to college to study for the ministry, he attended Randolph-Macon where he could focus on his studies full time.

Charles married in 1875. By 1890 when Vincent married, Charles and his wife, Emma, had welcomed their five children. Two years after Charles and Emma completed their family, Vincent and Daisy were beginning theirs. Although Charles didn't convince Vincent to join him in the family business, the two men ensured the continuation of their family bloodlines. Charles and Vincent were the only members of their sibling group who married and produced children.

Charles died at the age of fifty-four in 1907. Perhaps his health deteriorated from the physical exertion and emotional stress of being the only son working alongside his father, uncles, and cousins, to rebuild the family business after the American Civil War. After Charles died, the family business declined until all that remained were assets such as buildings and land.

During the final years of Vincent's career, the family remained in Virginia and moved several more times. From Richmond, they moved back to Norfolk, to Petersburg, then Richmond again, where Vincent tried (without success) to unite two congregations. He then went to Sanston, then to Greenwood (near Richmond), his last charge.

From the time of his father's death in 1909 until the last dependant died in about 1933, Vincent was financially responsible for many unmarried or widowed women in his extended family. Family lore indicated there were thirteen spinster ladies. Attempting to count those women, I reached fourteen before I stopped. Perhaps someone else was caring for number fourteen. The surviving heirs eventually divided the remaining estate, and Vincent was able to use his portion to better fund Daisy's household budget.

When Vincent retired in 1933, he purchased a home on Moss Side Avenue in Richmond, where he and Daisy lived for the rest of their married life, and where Daisy lived during her widowhood. Daisy's Richmond home was filled with furniture they purchased, inherited, or that Vincent built, and her much-loved piano.

Daisy was a wonderful cook. She accumulated linens, flatware, crystal, and china to accommodate church group meals she served in her dining room and on card tables in her living room. She loved updating her kitchen with the latest tools and appliances; and was the first woman among her neighbors to own an electric cooking stove. Her generally frugal lifestyle allowed Daisy to save for those important major purchases.

What were her husband's preferences? Vincent insisted on formal meals. Daisy, apparently in agreement, served their meals in the dining room. Daisy used fine china, sterling silver, and cloth napkins every day. When paper napkins became available, Vincent believed they were only appropriate for picnics. Children who dined at Daisy's table during Vincent's lifetime waited impatiently to eat Daisy's wonderful food while Vincent read from the Bible, then offered lengthy prayers. In the southern tradition of the era, breakfasts were hot, formal meals; lunch (called dinner) was the big meal of the day. Supper, still served formally, was a lighter meal than dinner.

Daisy served separate courses that included such favorites as roast beef, fried chicken, fresh vegetables, and tiny biscuits. Desserts were homemade pies, cakes, and cookies. Cousin Sue remembers eating wonderful fruit pies at Grandma Daisy's table. Sue went on to say Grandma Daisy was frequently upset when animals ate the cherries off the tree in her yard before she could gather the fruit.

Proper attire was required in the dining room, according to both Vincent and Daisy. Vincent always wore a jacket during meals. In summer, he wore a lightweight mohair jacket at the table. Shirtsleeves were not, in his opinion, formal enough. Daisy fully dressed for the day before preparing breakfast. She wore dresses with modest jewelry, a touch of face powder, and arranged her hair neatly on top of her head. In a minister's home, visitors frequently arrived unannounced with pressing needs such as illness or death of a church member. Perhaps Daisy realized that being presentable early each morning helped alleviate stress. When an unexpected knock came at the door, she was ready to greet guests.

Daisy's preferences and skills supported her role as a minister's wife. She adored hats when they were the perfect accessory for a minister's well-attired wife. Daisy was musically talented. She played piano and organ, and maintained her musical skills to play for church functions. Her musical talent was a great asset in small churches. She also loved opera. In later life, when she had more time and money, she attended opera performances as often as possible.

Daisy took her children to her parents' home for summer visits. Getting to Dorchester County on the Eastern Shore of Maryland from mainland Virginia wasn't easy. In Norfolk, Daisy and her children boarded a steamer for the trip to Baltimore. Fortunately, Daisy's Uncle Henry and his family lived in Baltimore, and welcomed Daisy and her children for an overnight stay. The following day, Daisy and the children boarded the paddle wheeler, Emma Giles, bound for the dock near her father's Madison home.

Since family members might romanticize their home county, I searched for outside opinions of the Dorchester County, Maryland area in the early 1900s. Annie Oakley and her husband, Frank Butler, visited the area and found it so appealing for hunting and fishing, they purchased several properties in nearby Cambridge as early as 1912. The couple visited many cities in the United States and abroad; then chose to retire to a waterfront cottage in Dorchester County, where they lived for several years with their dog, Dave. Annie, Frank, and their frequent guests thought the area a sportsman's paradise.

In Madison, Daisy and her children enjoyed homegrown vegetables from her parents' garden, seafood from the bay, and the fresh air and relaxed atmosphere of country life. Daisy's children enjoyed relationships with the offspring of her childhood friends. The children played and helped with picking and canning chores, Daisy's father provided fabric from his general store, and Daisy and her mother created clothes for Daisy's family. Madison summers were productive, happy times for Daisy and her children.

Daisy's mother, Sarah, died of cancer in 1907. Close to her mother as the only child for many years, then receiving help from her mother as an adult, Daisy must have felt Sarah's loss keenly. The following year, Luther remarried. His second wife, Molly, was the widow of Luther's cousin and someone Daisy knew well. Molly continued to welcome Daisy and her children for summer visits.

Daisy's brother, Oden, married in 1915. His first wife was apparently not equipped for marriage or motherhood. Daisy's first nephew, Oden Martin, was born and died in April 1916. A second son, Martin Luther, was born in March 1917 and died in July 1918. Poor mothering skills contributed to the boys' failure to thrive. Unsurprisingly, Oden's first marriage ended in divorce. Oden married Irma in 1926 who blessed him with a daughter, Sarah, who was born in June 1937. Daisy was delighted to have a niece.

Daisy's father, Luther, passed away in March 1918. Having outlived his second wife who died in 1912, his heirs were his two children. Following his death, Daisy and Oden inherited the family land. With no plans to return to live in Madison, Daisy and Oden initially kept property where valuable timber grew for continued income. They arranged to have the timber cut and sold, then divided the profits. Eventually, they sold all the family property, and shared equally in the proceeds.

Oden's daughter, Sarah, provided evidence of Daisy's relationship with her only sibling in the form of photographs of a 1931 family vacation. Daisy and Vincent joined Oden for a multi-generational gathering at Oden's summerhouse in Round Bay, Maryland, on the Severn River. One photograph taken during that vacation captures Daisy wearing a pretty dress and fancy shoes. Daisy, with a beaming smile, is holding her nine-month-old granddaughter, Sue, with grandchildren Grace Virginia and my father, John, nearby. She greatly enjoyed relaxing times with her extended family.

Vincent had skills and interests beyond his role as a pastor. He was ambidextrous and was said to write and use tools equally well with both hands; a skill developed unwillingly. Vincent was born left-handed then forced by his schoolmaster to write with his right hand. Even though he chose not to work in the family plumbing and metalworking business, he received training in those skills. He was able to accomplish repairs and building tasks in the homes he and Daisy occupied, avoiding the cost of paying others for such work.

Beyond saving money, Vincent found working with his hands relaxing. He collected tools of all kinds, and built a large chest for transporting them during their frequent moves. During my father's childhood, Vincent spent time developing his woodworking skills. On evenings when he did not have church obligations, Vincent went to the basement after dinner, and worked until bedtime creating objects from wood. Vincent allowed John to watch and participate, to his grandson's delight.

One of Vincent's celebrated woodworking projects was the design and construction of portable bookcases. Vincent's sermon preparation required hours of study supported by a large library of religious reference books. He enjoyed reading novels, when time permitted, and had a large collection of those as well. When reassigned, Vincent simply closed the bookcases and moved them. Daisy was pleased to avoid repeatedly packing and unpacking books.

Their children attended public schools then colleges such as Longwood, Blackstone, the University of Virginia, Columbia University in New York, Farmville State Teacher's College, the Virginia Mechanic's Institute, and Harrisonburg State Teacher's College. They sought gainful employment as soon as they could to help supplement the family income. Vincent Jr. was working as a bookkeeper for a life insurance company when he was eighteen. He attended business school, developed typing and shorthand skills, and secured a position as Secretary to the President of a railroad, a steppingstone to better jobs.

The first child of Daisy and Vincent's five children to marry was my grandmother, Marie. Since this book contains a chapter dedicated to Marie, you know her story.

In 1917, Vincent Jr. married the daughter of a North Carolina attorney and judge. Vincent Jr.'s career had taken him to Greensboro, North Carolina, where beautiful Eugenia lived with her prominent family. By all accounts, he was instantly smitten. Vincent Jr.'s World War I registration, completed shortly before his marriage, stated he was a railroad cashier at the time. Their son, another Vincent, was born in 1918 in Greensboro. They moved to Richmond in the 1920s, and then moved back to North Carolina. They were again living in Richmond in 1936 when their son was attending Richmond College. Daisy had opportunities to adore her first grandchild.

Daisy's second grandson, John, was born on Saint Patrick's Day of 1920 in Portsmouth, Virginia, where Daisy and Vincent were living at the time. Daisy created hand-made baby clothes for John as she did for her other grandchildren. I know the outfits she made for John were a well-remembered gift because of the phrase, "Grandma Daisy made his little outfit" penned on the border of John's toddler photograph.

Beyond clothing construction, Daisy could knit and crochet. She completed dozens of floral-centered wool needlepoint canvasses that covered chair seats and pillows in her home. Pillows I inherited still grace family sofas, so they are durable as well as beautiful. When Daisy's niece, Sarah, visited her in the spring of 1949, Daisy taught Sarah to crochet. Although Cousin Sarah says she didn't pursue the skill of crochet work, she has fond memories of her aunt Daisy spending the time with her, and Daisy's desire to pass along a craft she enjoyed. Daisy's daughters inherited her needle skills. My grandmother and her sister, Grace, completed many lovely needlepoint and crewel embroidery projects.

In addition to their own children and grandchildren, Vincent continued to have responsibilities for his adult siblings. He was named to care for the estate of his bachelor brother, Victor, who had gone west in 1890 and never returned to Virginia. Victor passed away in 1918 of complications following surgery in Lewiston, Idaho. No family member attended Victor's funeral. Friends buried him in an unmarked grave, and shipped Victor's wilderness belongings to Daisy and Vincent's home. One item was a bearskin rug they used until it fell apart. Thankfully, it was long gone before my time.

Let's reflect for a moment on Daisy's life during the two difficult years of 1920 and 1921. By mid-1920, she had two grandsons. One little boy, age two, was living in North Carolina, and the other was living nearby in Portsmouth, Virginia. Three children still lived at home. Emma, twenty-three, was working as a public school teacher. Grace, eighteen, was continuing her education to become a teacher. Walter, the baby of the family, was thirteen.

Emma married in 1920 and gave birth to Daisy's third grandchild, Grace Virginia, in 1921. Six weeks after giving birth, Emma died of double pneumonia. Grace Virginia's father felt he wasn't equipped to raise a baby, so he handed the little girl over to Daisy. As you know from Marie's chapter, Marie soon moved back into Daisy's home with her son, John. The challenge of caring for two babies pushed aside Daisy's emotional stress of losing Emma. Daisy selflessly carried the mothering load while Marie taught school to help defray her parents' household expenses.

Grace Virginia has said Daisy and Vincent were the only functional parents she knew. She saw her own father only occasionally for the rest of his life. She and John told me their grandparents made them feel loved and welcome. If Daisy was overwhelmed, she didn't allow innocent children to know her feelings.

Daisy and Vincent's daughter, Grace, married in 1926. She and her husband, Toppy, enjoyed a long, happy marriage and had two sons, Toppy Jr. and Bob. The youngest of Daisy's five children, Walter, married Elizabeth in 1928 and had two children, Sue and Antoine, bringing Daisy's growing grandchild brood to seven. In 1930, Marie and John moved to an apartment, and Daisy continued caring for nine-year-old Grace Virginia. After forty years of marriage, Daisy, fifty-seven, faced another decade of mothering a grandchild. Vincent, seventy-three, continued pastoring until 1933 and died in 1937. Although Vincent did not live to see them, Daisy welcomed seven great-grandchildren, and two more were born after her death. Her nine great-grandchildren have populated successive generations.

Daisy endured change with grace. During Vincent's ministry, she repeatedly adjusted to new homes, new churches, new towns, and new people, while raising the children and grandchildren God placed in her care. Following four years as the wife of a retired preacher, her widowhood stretched over a quarter century.

Although she lived until 1965, Daisy never drove a car. After Vincent's death, a woman who helped Daisy with housekeeping also provided transportation assistance. Daisy's grandchildren and great-grandchildren remember Lucy, who came to Grandma Daisy's house almost daily. Lucy helped with cooking and cleaning, served the meals she and Daisy prepared, and drove Grandma Daisy wherever she needed to go. Lucy was a gem.

Daisy's activities stretched beyond her household and church. She participated in other arenas where she made choices and took stands. Cousin Sarah told me Daisy was a member of the Women's Christian Temperance Union. The WCTU, organized in the 1870s and still in existence today, has fought for many worthy causes. Women joined forces under the WCTU banner in an effort to obtain pledges of total abstinence from alcohol then later from tobacco and drugs as well. They reportedly began working to protect women from abuse when some states seldom prosecuted rape. They fought to make it illegal for young children to have sex. Working for reforms to benefit women, the organization fought to obtain the right for women to vote, for fair treatment of women with respect to property ownership, and began efforts to protect women and children from abuse. Vincent agreed with the fight for abstinence and supported the welfare of women. Daisy and the females she raised were politely assertive individuals who voiced their opinions and were active in varied sectors of society.

Daisy was fearless. Family members have said she was particularly animated and unafraid when protecting the rights of others. One of her grandchildren told me of an instance occurring when Daisy was a widow in her eighties. Living alone in her Richmond home in the 1950s, she was fond of her next-door neighbors who were preparing to be away on vacation. When they asked Daisy to keep an eye on their home during their absence, she was quick to agree. Upon seeing a man attempting to break into her neighbor's home, courageous little Daisy armed herself with a shovel for self-defense, and marched over to confront the man. The burglar, startled by the tiny woman poised to attack, fled.

I would have admired Daisy if she'd *only* been spiritual, worn great hats, and baked cookies. She also had a loving heart, a quick mind, musical gifts, and a courageous spirit. During my visits with her, I do not recall going to church with Grandma Daisy. Perhaps I wasn't in Richmond on a Sunday during her lifetime. A few years ago, I enjoyed worshiping with the members of Richmond's Centenary Methodist Church where Daisy and Vincent were members after his retirement in 1933, and where she remained a member until her death in 1965. When I arrived that Sunday, I stood in the doorway and wondered where Grandma Daisy sat with her husband, and where she sat in her widowhood. The specifics weren't important; I was thrilled to worship where Daisy had worshiped.

Daisy's parents named her well. Daisies, according to Celtic legend, were scattered upon the earth by the spirits of children who died in childbirth to cheer sorrowing parents. Daisy comforted many weeping parents in her role as a minister's wife.

One of Vincent's peers compared Vincent's ministry to the man mentioned in Matthew 13:46. When God called Vincent to preach, Vincent was like the man looking for fine pearls. When the pearl-seeker of scripture found a valuable pearl, he sold everything he had to purchase that pearl. Vincent gave up a more financially lucrative career in his family's business to respond to God's call. Perhaps Vincent's earthly pearl of great value was Daisy. She was a helpmate who facilitated his efforts to lead others to the superior prize of eternal security. I wasn't surprised to learn the derivation of the French word for daisy, "Marguerite," stems from a Greek word meaning "pearl." How fitting.

When I asked a pastor's wife to describe the challenges women in her role face, she said the need to be an encourager for her husband was more challenging than her direct interfaces with church members. She explained that pastors feel they need to exhibit all the Godly attributes they speak of on Sunday morning. Pastors are humans saved by grace, who (like me) will fall short, and need to forgive themselves as God forgives them.

As the scripture verse found in Proverbs 31:11 teaches us, a husband who puts his confidence in a capable wife will never be poor. Daisy, petite and youthful at seventeen, must have trusted God for the maturity and strength she needed for her busy married life. With little money, she spiritually and literally fed and clothed her family, and warmed three generations with her love.

Daisy taught me to stay the course. Spiritually strong women are not surprised when life delivers challenges. They overcome grief and difficult circumstances with grace. Even housed in petite bodies, they are brave. They stand firmly on the scriptural teaching that He who lives in them is greater than he who is in the world. They make the best of material shortages. They offer welcome and love to parentless children.

Did I find Daisy's prayer book or her Bible? No, I didn't. When I read the attributes of a spirit-filled person in Galatians 5:22, those words describe the woman I called Grandma Daisy. Her life illustrated love, joy, peace, patience, kindness, goodness, faithfulness, humility, and self-control. I need no further proof. She was a great spiritual role model for all who knew her.

Daisy is now enjoying something superior to the beautiful Christmas cookie she gave me when I was a child. Daisy has received the sweetest reward. She has joined Vincent and their Lord in paradise.

PART IV

Great, great-grandmothers

Hannah

1812 – 1886

Hannah, whose name means *grace* in Hebrew, descended from a family crowded with women named Hannah. To preserve our sanity, I'll identify these different women named Hannah by using their birth year along with their first name. Hannah (1812) is the woman whose life we'll explore in this chapter. She was my great, great-grandmother. You've read about her daughter, Julia, and her great-granddaughter, Grace.

My Hannah (1812) had a Virginia-born, great-grandmother named Hannah (1718). Hannah (1718) had a brother named John. John's daughter, Hannah (1736) married John Augustine Washington, the younger brother of President George Washington. Upon their marriage in 1756, John Augustine Washington took my kinswoman, Hannah (1736), to live at Mount Vernon where she gave birth to their first two daughters. John Augustine's brother, George, brought his wife, Martha, to live at Mount Vernon in the summer of 1759. At about that time, John Augustine Washington and Hannah (1736) moved to her nearby ancestral estate. Hannah (1736) was returning to the home of her birth where she had lived until her marriage three years earlier.

John Augustine Washington's move to our ancestral family home was prompted by the illness of his wife's father who probably needed help running his plantation. Hannah (1736) and John Augustine Washington, along with their children, joined the household of Hannah's father, John, and his third wife, Mildred Washington (Widow Seaton). Mildred was a first cousin of John Augustine Washington and George Washington. At her father's death in December 1760, Hannah (1736) inherited our family estate, thereafter known as one of the Washington family homes. My ancestors created an interesting, tangled intermarriage web indeed.

Do I have a drop of Washington blood in my veins? No. I descend from Hannah (1718), the aunt of the woman who married a Washington. My direct ancestors and people of the Washington family did attend, however, the same church services, weddings, and dinners. They exchanged overnight visits and corresponded by letter.

Who might have gathered around the tea tables of our ancestral women? Even better, who might have shared a church pew with our kinswomen? Imagine how much more pertinent the study of history will be for children and grandchildren if they closely relate events to their ancestors. Learning where our ancestors were living, and what they were doing, will make each historic era more meaningful. When members of our families intermarry with a famous family, books abound. By describing the home of George Washington's brother, authors and historians provide marvelous descriptions of my own family seat in Virginia.

Now let's quickly walk across a *people bridge* from my ancestral Washington-connected family of Virginia to my own line. Virginia-born Hannah (1718) married her second husband, Adam, in 1742. Hannah and Adam had a daughter, Hannah (1746). Robert, a Virginia neighbor, married Hannah (1746) and she had a son named John. The family then moved to North Carolina where John met and married Ann. John and Ann are the parents of Hannah (1812), the subject woman of this chapter.

If you are fortunate, other people with whom you share DNA will research your family history and share their findings with you. As mentioned in Julia's chapter, two cousins of my mother's generation, Frances and Katharine, spent several years researching and documenting the history of four major family lines. Frances and Katharine wrote extensively about the people of this chapter in their self-published book, *The Points of Our Compass: Surveying Our Ancestors*. For brevity, I'll mention Cousin Frances when referring to their work.

My great, great-grandmother, Hannah, was born on November 30, 1812 in Randolph County, North Carolina. She was the youngest of nine children in John and Ann's family. Cousin Frances's book tells us Hannah was probably born in the town of Asheborough (now Asheboro), the county seat of Randolph County, where the family lived for many years. From the time Hannah was about three years old until after she married and gave birth to her first two children, she lived on Back Creek Plantation near Asheboro.

Hannah's father, John, apparently had a plan for his family that was well underway before Hannah was born. Since it appears his careful planning and hard work created the life Hannah knew as a child and young woman, we'll look briefly at John's activities.

In 1793, John purchased two lots in the town of Asheborough where he built a general store and a house. About 1796, he married Ann whose father owned a nearby tavern. John initially supported his family as proprietor of his general store. Beginning in about 1803, he gradually purchased over nine hundred acres of land then moved his family to Back Creek Plantation.

Most plantation owners of substantial means hired tutors for their children. Hannah's parents followed suit and went a step further. John and Ann constructed a building on their plantation to serve as a school and church for their family and for people from surrounding plantations. Hannah must have had a small library at her disposal; her father's 1835 will mentioned fifty-three books.

Cousin Frances believed Hannah's life was probably typical of life on an affluent plantation in her day. One of Hannah's brothers became a Methodist minister and her mother's obituary points to Methodism, so Hannah and her siblings worshiped as Methodists in their youth.

As the last child born to her parents, Hannah was only two when her siblings began to marry and leave home. Before Hannah's arrival, her mother had given birth to eight children: Nancy, Cynthia, Martha, Frances, John, James, William, and Eliza. Let's look briefly at Hannah's siblings in the order of their birth. Why look at the lives of a woman's siblings? Hannah learned to deal with the difficult moments of her adult life by observing her mother's relationships with Hannah's siblings as children and as adults. The lives of her successful (and not-so-successful) older brothers and sisters informed Hannah's choices of husband and lifestyle. Now, in their birth order, we'll look at Hannah's siblings and learn a bit about their lives.

Nancy married in 1816. Hannah's eldest sister moved to Tennessee in 1830, then moved to Mississippi in 1842, and had ten children. It is unlikely Hannah saw Nancy after she left North Carolina.

Cynthia married in 1814 and had a daughter. Widowed, she remarried and moved to Tennessee, leaving Emily, her daughter by her first marriage, with her parents. Since Hannah was only three years old when Emily was born, they grew up more as sisters than as aunt and niece. Hannah observed her mother raising a grandchild as her own child.

In 1818, Martha married a brother of her sister Nancy's husband. Martha died eleven years later leaving five children. Her widower moved west so Hannah's family lost Martha and lost touch with her children.

Frances married in 1825. She had no children and her husband died 1837. She then married a man who had a son by his previous marriage. With her second husband, she had two daughters and lived near Hannah's family.

John Jr. never married and died on August 31, 1827. His obituary stated he died at Fayetteville, North Carolina, in the twenty-fifth year of his life. His death must have been a shock to the family. When John died, the women were already wearing mourning clothes from the death of another brother about whom you'll read as the family story unfolds.

James reportedly trained as a physician. There is no evidence he practiced as a doctor or worked diligently at any career. At thirty-two, he married Eliza, then fifteen. They lived on land his father had given him near his parents' plantation home. To pay his debts, James began selling parts of the land he had received from his father. He also sold furniture, livestock (even his wife's milk cow), and household items including their oven, skillet and water bucket. The careful wording of his father's will indicates James was not trusted. Ann inherited the family plantation at her husband's death. At Ann's death, James was heir to the land, but "To be held by him, his Heirs, and assigns forever." In other words, James was *not* to sell the land.

After her husband's death, Ann tried to help James. She tried to bail her son out of debt and nearly lost her plantation in the process. James died in 1846 at the age of forty leaving a twenty-three-year-old wife and three young daughters. Ann continued to pay off her son's debts and supported her daughter-in-law and grandchildren.

Four years after James died, the 1850 census record indicates his wife and daughters were still living on the plantation next door to Ann. When James's difficulties were occurring, Hannah was married to Sampson who was overseer of her father's plantation. Sampson must have thought James's behavior deplorable. A hard-working, self-made man, Sampson could have helped James financially, although he probably refused to do so because James was untrustworthy and lazy. Did James produce anything good in his lifetime? What happened to his descendants? James's children apparently learned better character traits from their mother, Eliza, or their grandmother, Ann.

We'll look at the second daughter of James and Eliza as an example. Eugenia married, had one son, and then was widowed at the age of nineteen. Eugenia worked diligently to support herself and her child. She first taught school, and then was a respected postmistress for forty-two years. She was a beautiful brunette with stately carriage. Her son produced five children whose lives reflect talent, intelligence, and drive. Their descendants are numerous.

William, much to their mother's delight, became a Methodist minister. At the age of nineteen, he began preaching *on trial* as second man to the Mecklenburg Circuit in North Carolina. His next assignment was the Black River Circuit in Sampson County, North Carolina where he was preaching to two hundred and fifty white and ninety-eight black church members until five weeks before his death in June 1827. After William's death, his parents received a beautiful tribute published about him in a Christian journal. Hannah's mother had lost a child whose life seemed so promising. At fourteen, Hannah lost the brother closest to her age and whose profession she probably admired.

After looking at the lives of Hannah's brothers, a sad reality comes to mind. After fathering nine children, three of whom were sons, Hannah's father died without a male grandson to carry on the family name. The only one of his three sons to marry and have children had three daughters. Since James outlived his father, perhaps John died hoping a grandson might be born to him posthumously. It wasn't to be.

Eliza married a doctor in 1827 and the young couple moved to Tennessee. Eleven months later, in November 1828, Eliza died there following childbirth. Relatives with whom she stayed in Tennessee wrote to Eliza's father following her death. The letter, written in Rutherford

County, Tennessee, on December 3, 1828, appears verbatim in Cousin Frances's book. It clearly communicates the difficulties women faced in the early 1800s so I thought it worthy of inclusion. The author, Nancy, was Eliza's second cousin. It reads:

> *Dear Sir:*
>
> *I take this opportunity of informing you of the last remains of your daughter Eliza who married Dr. William, and to describe the treatment of the doctor to your daughter after they came to my house in Rutherford County some time early in October.*
>
> *She was in a low state of health when they came to my house, and she continued so until she had her infant. She told me repeatedly that she felt quite smart, except weakness. In two weeks after the birth, one of her breasts began to rise, and there was no prevention for it. She frequently told me if it were not for her breast, she would soon get well. About twenty-four hours before her death she told me that her breast had got quite easy, did not hurt her at all, and she was taken with spasms and shortness of breath; which continued for eight or ten hours and terminated her existence.*
>
> *In all her afflictions from first sight, husband, mother, father or any other person could not have paid more respect and attention to a companion or any relative than Doctor William did to his wife since he arrived at my house. It was remarked by every person that saw them that he was so affectionate to her, more so than common. With respect, Nancy*

An additional notation illustrates the mental condition of Eliza's husband:

> *The Doctor talks of taking his wife to North Carolina next fall. You would do him very great favor to persuade him to desist in so great an undertaking. I think by that time, it would be impossible; for I think mortification took place before she died. When we put her in her coffin, she smelt very disagreeable.*

Imagine Hannah's family reading the heartbreaking account of Eliza's death. The loss of Eliza at this particular time must have been a devastating blow for Hannah and the rest of their family. The year before, Hannah's brothers William and John had died. Eliza's infant child must have died as well because we see no mention of the baby in the letter about Eliza's death. Seven years later, in their father's 1835 will, he asked Hannah and her siblings to make every effort to retrieve money he had given Eliza and, once retrieved, to divide the money among John's heirs. Surely if Eliza's child survived, John would have wanted his grandchild to have the money.

Before I go on with Hannah's person life, let's look for a bit longer at her mother who, I believe, taught Hannah to be a strong, principled person. Ann's thirty-nine year marriage produced nine children. Ann outlived her husband and six of her nine children. When Ann stood at her husband's graveside in her early sixties, she must have anticipated joining him soon rather than living another quarter century. Several children and grandchildren depended upon Ann's help during her widowhood. God had a purpose in extending her life. Ann's exemplary strength and grace taught Hannah to endure the future difficulties of her life as a wife, parent and grandparent.

Ann, according to her obituary, worshiped as a Methodist Christian for over forty years. She was opposed to slavery. Her husband, who apparently didn't agree with her viewpoint, specifically "lends" slaves to her in his will to preclude her from freeing them. Consequently, Ann cared for five elderly African-American former slaves until her death.

Cousin Frances learned Ann kept silk worms, made silk thread, and used the thread to weave silk cloth. Ann made a one-yard-square shawl with six-inch macramé fringe that was in perfect condition one hundred and twenty-five years after her death. People who saw the shawl considered the quality of the cloth and the workmanship of the finished shawl excellent. Although I don't know whether Hannah learned those skills from her mother, we know Hannah had a productive, creative mother after whom to pattern her life. Ann's last residence was a little log cabin near the main house where Hannah and her family were living in northern Randolph County. Ann's opposition to slavery became Hannah's stance as well.

In 1855, the curious notation "of God" follows Ann's name on a roster documenting school and church attendance. Cousin Frances surmises the notation meant Ann was a member of the church that met in the school building. Ann's name on the attendance roster in 1856 included an annotation that she was a widow and infirm. The Randolph County 1860 Mortality Schedule includes the entry: "Ann, widow, age eighty-eight years, died in March 1860 from pneumonia after a five-day illness." In 1860, the church attendance roster noted Ann was infirm then deceased. Cousin Frances assumed her burial location, although not confirmed, to be next to the grave of her husband on Back Creek Plantation.

Hannah spent her youth on Back Creek Plantation where she worshiped and attended school under the influence of her parents and siblings. In about 1830, the course of her life changed when her father hired an eighteen-year-old Quaker man as overseer of their plantation. Sampson, descended from Scots mentioned in Julia's chapter, sought employment because his father had passed away leaving no land for his son. Sampson could read, write, and had a keen mind for managing business details.

Apparently, Hannah became acquainted with her father's overseer. Perhaps she served his meals at her parents' table. On May 22, 1834, Sampson and Hannah (both twenty-two) signed their Randolph County marriage bond. If Hannah's father felt his daughter married beneath her place in society, there is no hint of such feelings in legal documents. John wisely made Sampson an administrator of his estate. During those difficult times when Hannah's only surviving brother, James, was draining his mother's resources and selling land, Sampson purchased some of the plantation himself, thus keeping the land in the family for his wife and their heirs.

Photographs and physical descriptions show Sampson and Hannah to be an odd-looking couple. He was over six feet tall and, in a photograph located by Cousin Frances, Sampson appears quite stocky with a full face, blue eyes, and thick, curly, dark hair. Sampson posed for his photograph wearing a dark suit with a vest, white shirt, and bow tie. His left hand, shown gripping his left jacket lapel, is broad and appears strong.

Hannah was a petite, brown-eyed, brown-haired woman. In a photograph taken when she was about forty years old, Hannah wore a simple, dark dress with full sleeves, a white collar, and white cuffs. An edge of white lace visible around her head indicates she was wearing a cap or small bonnet held on by a ribbon tied under her chin. Her face was long and thin; her hands were small and slender.

Sampson was a productive husband determined to provide well for his family. Beyond farming Hannah's family plantation for his mother-in-law,

Sampson purchased at least two hundred acres of land, and Randolph County appointed him overseer of a nearby road. In January 1838, Sampson placed an ad in a local newspaper announcing his purchase of a tavern on the corner of Courthouse Square in Asheborough. The ad described large, pleasant rooms, dry stables and satisfying meals. Sampson invested his tavern profits in land. He paid $150 for seven hundred acres of land along Back Creek in February 1838.

Hannah was building a family. During the first fifteen years of their marriage, Hannah had four daughters and three sons who her *Ma*. The 1850 census taken in Randolph County, North Carolina, provides a snapshot of their household: Sampson (thirty-seven), Hannah (thirty-seven), Virginia (fifteen), John (thirteen), Robert (eleven), Martisha (eight), Harriet (six), William (five), and Julia Jeanette (my great-grandmother, one). Between 1856 and 1858, Sampson moved his family to a farm called Vickery Place.

Sampson, Hannah, and their family (including two sons born since 1850) were living at Vickery Place when the 1860 census taker recorded their household. Their family members included Sampson (forty-seven), farming with sons John (twenty-two) and Robert (twenty). William (twelve) and Franklin (eight) were attending school; Hannah's youngest child, Edwin (five), was still at home. Females in the household include Hannah (forty-seven), keeping house; Virginia (twenty-four), a tutoress; Martisha (eighteen), a seamstress; and Julia (eleven), attending school.

What was happening in Hannah's spiritual life? During her adult life, she and her mother were instrumental in creating family memories of worship and education. They began a tradition of pulling the curtains closed on the Sabbath to (they said) shut out the distractions of the world. Behind those curtains, Hannah and Ann were unlawfully teaching African-American children and adults to read, using the Bible as their primary text.

During the American Civil War, three of Hannah and Sampson's five sons fought for the Confederacy. One son spent time in a Union prison camp; all three survived. A family story that gives me pause states the sons who went to war did not return home because Sampson had been such a stern taskmaster. Sampson, a hard-working self-made man, wanted his sons to be productive members of society. Perhaps Sampson overcompensated after watching the failure of Hannah's brother, James.

Unlike women whose sons left home to fulfill dreams, Hannah's separation from her sons and their families stemmed from her husband's harshness. We'll catch up with the boys and their stories as we continue through the years of Hannah's life. Did her mother, Ann, model strength when her own children moved far away? Perhaps Ann's example helped Hannah adjust to separations with some of her own children.

By August 1870, Sampson and Hannah had moved to Sumner Township, Guilford County, North Carolina. Sampson's desire to live in a thriving Quaker community drove him to locate his family near Centre Friends Meeting. The children in the home in 1870 were Julia (twenty), and sons Franklin (seventeen) and Edwin (thirteen) who were helping their father run the farm.

By 1880, Hannah and Sampson's hometown had been renamed Pleasant Garden. According to census records in 1880, Sampson and Hannah were both sixty-seven years old, he was working as a farmer, and she was keeping house. In Pleasant Garden, Hannah and her children gained reputations as intelligent people who expressed their faith by singing hymns of praise in church and at home. Neighbors often gathered in the home of Hannah's musical family where she *pitched* songs at the proper starting note.

Hannah's grown children began their own families. Her eldest daughter, Virginia (called Sister by her siblings), might have appeared to be heading for spinsterhood. At age 34, however, she went to a family wedding and met a man whom she married in 1868. After having given birth to five daughters (one of whom died at age two), Virginia's husband died of tuberculosis in 1880. Virginia died of tuberculosis in 1881.

At the time of Virginia's death, the four surviving daughters (two were twins) ranged in age from seven to twelve. One of the twins, Mary Julia, remained with her grandparents, Hannah and Sampson. The three other girls lived in the nearby homes of aunts and uncles and returned often to visit their sister. Hannah had learned about raising grandchildren from her mother.

Hannah's second child, John (named for Hannah's father), married Martha in 1867. Hannah knew their first child, her granddaughter Cora Lee, who was born in 1868. Before the birth of their second child (James, born in 1875), John had moved his family from North Carolina to Illinois where they stayed about fourteen years. John then moved his family to Fort Worth, Texas where he spent the rest of his life and died in 1911. John and Martha added seven grandchildren to Hannah's descendants: Cora Lee, James, Ernest, Mamie, Winifred, Carrie, and William.

Perhaps letters Hannah received from her granddaughter, Cora Lee, lifted some of the sadness at not seeing John's children. Even after Hannah's death, Cora Lee wrote faithfully to her aunt, Julia, and to other female relatives. How proud Hannah must have been of Cora Lee who spent her career as a speech teacher in public schools and a private studio, having continued her post-graduate studies at universities in Texas and California through the years. Cora Lee's younger brother, James, was musically talented as was their sister, Winifred, who helped support her family by giving music lessons. Musical talent continues to be a noticeable trait generation after generation among Hannah's descendants.

Born in 1839, Hannah's third child, Robert, exemplified the family's musical talent. Cousin Frances's book includes a newspaper article published in 1932. The article quoted an eighty-one-year-old man who remembered seeing Robert and his sisters gather on meeting days at Centre Friends Meeting to sing for appreciative audiences. The man recalled first hearing *Dixie* sung when Robert, then a young officer in Confederate uniform, received a request for the song. Robert, the man reported, stood in the rear of his buggy and thrilling the crowd with his splendid baritone voice.

Robert married the first time in North Carolina and had a daughter named Versa then his wife died. Although his first wife's name, the date of their marriage, and the date of his wife's death have not been found, they must have married shortly before Robert's conscription into the Confederate Army on August 1, 1862. Entering his regiment as a First Lieutenant, he received a promotion to Captain on January 1, 1863. Captured and imprisoned in Georgia, he eventually signed the oath of allegiance, and joined the Union Army. He wrote that the Union enlistment was against his will although his sympathy had been with the Union cause all along. His feelings likely stemmed from his mother's abolitionist leanings. Having sons fighting in the war must have kept Hannah on her knees in prayer.

Robert married his second wife, Belle, in 1873 after arriving in Illinois following the war. Robert and Belle had no children though they helped raise Belle's younger sister. Robert followed a career appropriate to coming from a musical family by composing, singing, teaching music, and owning a music store. Belle worked with Robert in the store to allow him more time for teaching and composing.

Once in Illinois, Robert did not see his parents or siblings again. In 1886, upon hearing of the death of his mother, Hannah, he reportedly went into seclusion and wrote the tune and lyrics to a song, "My Old Cottage Home," about his early childhood years at the old family home at Back Creek Plantation. The song mentions it had been many years since he knelt with his dear ones around the old hearth and heard his mother's sweet prayers. He later said the words came to him as fast as he could write them down, and the music followed a few minutes later. Thanks to Robert, we know Hannah prayed aloud with her children.

Apparently, Robert's talents were widely known. Our family received a story about a passenger on a train passing through Murphysboro, Illinois where Robert and Belle's music store was located. The passenger, who had known Robert in North Carolina, asked one of the townsmen if they knew Robert. The man answered, "Yes, everybody knows Bob, the sweet singer," and went on to say Bob was old yet he continued operating his music store and was still a great singer. Robert also published a ninety-six page book of instruction for singing classes. Robert faced health issues including asthma and rheumatism in his declining years, and Belle struggled with poor eyesight. They kept working as long as possible, then signed over their store and home to their Methodist church, and lived out their lives in a Methodist retirement home.

Mary Martisha, called "Tishie," was Hannah and Sampson's fourth child and called the liveliest and most fun-filled child of her generation. Perhaps she needed a positive attitude to help her live happily for ninety years. Born in 1841, Tishie married a Methodist minister with whom she had one daughter, Nannie, born in 1871. Tishie's husband died of pneumonia on March 16, 1885. After her first husband's death, Tishie remarried but had no more children. Her second marriage took her to her new husband's lovely one-story Victorian cottage in Carthage, a lovely North Carolina town. Her second husband died in August 1894 leaving Tishie alone again. As she grew older, she lived for a time in a boarding house near her Methodist church where she worshiped as long as she was able. For extra money, Tishie sold divinity candy she made over a hot plate in her room. When she could no longer manage on her own, two grandchildren placed her in the Confederate Home in Fayetteville, North Carolina where she lived out her ninety years.

Harriet "Hattie" Louise, born in 1846, was the fifth child in Hannah's household. She married John, a Confederate veteran and minister's son, in 1867. After their marriage, Hattie and John lived with John's parents in Franklinton, North Carolina. Hattie made and sold crocheted items to purchase a parlor organ. She had grown up singing with her siblings, particularly Robert and Tishie, so having her own musical instrument was important to her.

On September 9, 1878, Hannah and Sampson transferred a parcel of land valued at one hundred dollars to Hattie. Perhaps Hattie sold or leased the land to aid her finances. She and her husband continued living in Greensboro, a city not far from Sampson and Hannah's country home. Hattie was a congenial, popular woman. Cousin Frances remembered Aunt Hattie as "a tall, thin, somewhat solemn woman with gracious dignity" whose impeccable table settings included finger bowls. Hattie outlived her husband by a decade and died on December 27, 1918.

Hannah's brunette, brown-eyed looks crop up repeatedly in her grandchildren. The musical tradition continued into successive generations as well. One of Hattie's granddaughters (Hannah's great-granddaughter) became an accomplished organist who was much in demand at Methodist and Presbyterian churches for weddings and special occasions.

Hannah and Sampson's sixth child was their son, William, who served in the Confederacy during the American Civil War then went west. He lived in Illinois and worked as a salesman. After arriving in Illinois, he married a woman named May and had one son, Willie. William was alive in 1910 but died before 1920 when the census noted his wife, May, was a widow living with their married son, Willie. Hannah must have missed William and wished to meet his wife and her grandson.

Julia Jeanette, my great-grandmother (whose story you read in her chapter), was Hannah and Sampson's seventh child. Hannah's relationship with Julia mirrored Hannah's own relationship with her mother, Ann, since they were both last daughters. Julia was born when Hannah was thirty-seven years old. After Julia's marriage, Julia lived in close proximity to Hannah for the rest of Hannah's life (another thirty-seven years). Hannah lived long enough to witness the birth of all four of Julia's children, including the last daughter born in 1885 and named Hannah Mozelle.

Hannah's eighth child was Franklin "Frank" Fillmore born in 1851. A five-foot, eight-inch tall man weighing about one hundred and sixty-five pounds, Frank married his neighbor Martha, a sister of Julia's husband. Frank and Martha had ten beautiful, charming, musical children. Frank and Martha's son, Vernon, acclaimed as the most handsome and musical of the ten children, died in 1903 at the age of twenty-four. Frank died in 1910.

Hannah and Sampson's last child was Edwin Columbus, born on April 3, 1855. Called Eddy, he lived a fruitful eighty-seven years. In 1888 (about two years after Hannah passed away) Eddy was admitted to the North Carolina Methodist Conference then spent twenty-three years as a pastor and twelve years as an evangelist. You read about his career in Julia's chapter. Did his formal training provide the most benefit or was it Hannah's example of prayer and Bible study that prepared him for a spiritual calling? Even before Eddy entered the ministry, he was diligent at other tasks. He grew up working on his father's farm and at twenty-two, received an appointment as overseer of a public road.

In 1886, Hannah's name appeared on a list of individuals who owned land in Guilford County, North Carolina. According to the list, Hannah owned 118.5 acres of land. Did she purchase her land from the proceeds of the sale of her family plantation in Randolph County? I assume that was what had occurred. It is puzzling that Sampson's name does not appear among people who owned land in their county in 1886. Sampson might have owned land in another county or he lost land during the financial downturn of war and reconstruction. Whatever the circumstances, it seems odd that Hannah owned property when her husband's name is conspicuously absent.

On July 26, 1886, Hannah died. She had been a loving, giving daughter to her parents, an obedient wife, and a dedicated mother to her children. Thankfully, in addition to teaching her children practical skills, Hannah encouraged them to be spiritually strong and to develop musical talent to praise God. Her descendants continue to bless their churches and communities spiritually and musically.

Hannah's four daughters received excellent educations for women of their day. They married well and produced fifteen children. Although Virginia died at forty-six in 1881 leaving four orphaned daughters, the other three daughters lived to the ages of seventy-two, eighty-six, and ninety. Hannah's sons fathered nineteen children, lived long, productive lives, and all reached at least the age of seventy. Sadly, when her older sons left North Carolina, she probably did not see them again or meet their wives or children. Perhaps she did all a mother in her position could do; she wrote to them and prayed for them.

For fifty-two years, Hannah was married to a Christian man who provided well for his family and demanded no less of himself than he asked of others. Hannah, to my knowledge, never complained of Sampson's treatment of her or of their daughters. When Sampson's reportedly harsh upbringing caused their sons to go west, Hannah lost time with her sons and their families. Perhaps having seen her sons become productive men, she found solace in realizing Sampson's lessons, although perhaps too harshly applied, helped their sons succeed.

Sampson lived nine years after Hannah died. If Sampson had predeceased Hannah, perhaps her sons who moved away would have returned to live close to Hannah, or she might have moved to live with one of her sons. We'll never know. Spiritually strong women focus on children who are close and find ways to be influential in the lives of children who are far away. They don't waste time complaining about distance but work creatively to close the gap. We can, with God's help, rise above life's difficulties and make the best of situations we're powerless to change.

Hannah loved God and people. She abhorred slavery. Beyond directly sharing God's message and love with others during our own lifetime, women have an opportunity to raise children to be productive citizens equipped to go forward and spread their faith.

Denomination wasn't the focus of Hannah's spiritual life. Her parents probably started life as Episcopalians then became Methodist. One of Hannah's brothers became a Methodist minister, as did one of Hannah's sons. When Hannah and Sampson moved to Guilford County, North Carolina, they appear to have returned to Sampson's Quaker roots. They lived about two miles from a Quaker meeting where they worshiped and where their remains now rest. Those descendants who knew Hannah spoke of her as a Godly, praying woman. Women who love God, study his Word, pray, and love others leave a lasting, positive mark.

Based on the evidence of her faith, Hannah is now in God's heavenly house where church membership certificates are meaningless. To enter heaven, our names must appear in the *Lamb's Book of Life*. Is your name there now? Are you certain? If not, talk with a trusted, Bible-believing individual or pastor. You'll want to spend eternity in the presence of God, where I pray you'll find your own spiritually strong ancestral women.

You won't want to miss meeting Hannah. When she isn't busy giving the heavenly choir the proper starting note, she'll be the woman sitting by the hearth praying aloud.

Tabitha

1829 – 1912

My grandmother, Lena, spoke often of her grandmother, Tabitha. I knew of this great woman's character long before I saw her face. After years of searching for a photograph, I found Tabitha's descendants, Cousins Mike and Kaye. They referred me to Cousin Ed who drove from Florida to North Carolina, scanned an undated photograph of Tabitha with her husband, Sanders, and sent it to me via e-mail. I have the best cousins.

Based on Sanders's date of death and his mature appearance, they posed for the photograph in about 1863. Sanders (about thirty-eight) appears stocky and solemn with broad cheekbones, an aquiline nose, dark hair, a slightly receding hairline, and a full, dark beard. His attire is a dark Sunday suit with a vest, a light-colored shirt, and a bow tie. With his hands resting his lap, he grips his left wrist with his strong-looking right hand.

Dark-haired with an oval face, Tabitha has wide-set eyes under straight, dark eyebrows. Her lips appear pretty and plump, even in a serious pose, so perhaps they were her best feature when she smiled. She wore her hair parted in the middle and pulled back, braided or twisted into coils, then tied on the sides of her head with ribbons. Her outfit includes an embellished, medium-colored jacket with wide sleeves over a white or light-colored blouse. Dark lace gloves cover her folded hands.

This dear woman is chief among the strong women I've found in all my ancestral lines. Tabitha experienced repeated tragedies, yet she mustered the energy to make life easier for those around her. We'll begin Tabitha's story in her early years.

Tabitha Jane was the fourth child and only daughter of her parents, Alexander and Rebecca. Born on August 23, 1829, her first name honored her maternal grandmother; her middle name echoed the name of her maternal great, great-grandmother, great aunt, and aunt. Tabitha's first name means *gazelle*. Her parents probably didn't realize that beyond honoring the child's grandmother, they were naming their daughter after a small antelope known for its graceful speed. The name Jane, a variation of the name Joan, is the feminine form of John and means *God is good*.

Tabitha, who answered to her first name, endured variations in spelling and shortened forms that typically plague difficult names. Alternately referred to as Tobitha, Tabatha, Tabbitha, Abitha, and Byther, we'll call her Tabitha. According to my grandmother, proper pronunciation of Tabitha's name places the accent on the second syllable (Ta-bi'-tha).

Tabitha (at five and at fifteen years of age, respectively) appears with her parents in 1830 and 1840 census records in Guilford County, North Carolina. Alexander was a farmer and served as clerk of the county court. He was an educated man, and census records indicate his children attended school as well. Several schools operated near Tabitha's home during her childhood, but I haven't found proof of Tabitha's formal education. Whether schooled at home or in a formal setting, she was educated.

In addition to the three brothers born before Tabitha, she had three younger brothers as well. As the only daughter, she and her mother must have been busy keeping house, sewing, and cooking for themselves and seven male members of their household.

On November 29, 1849, Tabitha married. Her husband, George Sanders, used his middle name as his call name. Sanders's male line descends from John, an Irish immigrant who arrived in America about 1737. Initially settling in Pennsylvania, John moved his growing family to North Carolina in about 1753 when his son, Robert, was a young boy.

In North Carolina, John acquired enough land to divide among his numerous descendants for several generations. Although no marker remains, a record of John's burial in a Presbyterian cemetery provides a clue about the family denominational choice during their early North Carolina years. John's son, Robert, grew to manhood and had a family that included a son named Samuel, born in 1781. Samuel married and worked as a farmer to support eleven children in Guilford County, North Carolina.

Sanders, the tenth of Samuel's eleven children, became an orphan at the age of nine in 1834. Unlike some orphaned children who begin adult life destitute, Sanders inherited land from his father and had siblings who functioned as surrogate parents. Sanders's eldest brother was fifteen years his senior and had received instructions in their father's will to ensure proper care and education for Sanders and other younger siblings.

To use a North Carolina term, Tabitha married a man from *good stock*. His ancestral fathers acquired and farmed land and Sanders was industrious as well. According to the census of 1850, he had real estate (perhaps inherited) valued at six hundred dollars. Ten years later, the 1860 census notes he had real estate valued at two thousand dollars plus a personal estate of two thousand, eight hundred dollars. During the first decade of their marriage, Sanders tripled his land holdings and established what appeared to be a growing personal estate. Between 1850 and 1860, Tabitha gave birth to four of her five children: Julius Alfred Cicero, Emma Florence, Franklin Alexander, and my great-grandmother, Isabel Anne (Annie Belle).

Life as a housewife in the 1850s was physically demanding. Wartime shortages of the 1860s made a woman's effort to feed her family even more difficult. Merely reading an 1863 description of preparing hog meat for storage using little salt (a scarce commodity during the American Civil War) made me tired. The directions required cutting up the freshly killed meat, dipping it in boiling brine for one to three minutes, cooling, packing the meat in barrels of cold brine, drawing off the brine in forty-eight hours, re-boiling the brine, covering the meat with brine again, allowing it to stand eight to ten days, then hanging the meat to be smoked. The recipe stated that by re-boiling the brine, one could reuse it for several killings if properly skimmed. Of course, reading such a recipe made me wonder how many deaths resulted from improperly prepared meat.

Tabitha's husband was a farmer who grew vegetables, cotton, and tobacco. They had cows for dairy products and meat, chickens for eggs and meat, and raised hogs for butchering. In an era before refrigeration, gathering raw materials and cooking large meals filled much of each day leaving little time for other pursuits. Yet Tabitha found time to sew, nurse babies, teach her children, and attend church. The 1850 census indicates she had help with her household chores in the early years of her marriage. A mulatto woman and her two children were living in Tabitha's household in 1850. By 1860, however, that woman was gone.

Because his wife and five young children depended on him for their livelihood, Sanders did not join the Confederate forces early in the American Civil War. With Sanders safely at home, names of other family members inhabited Tabitha's prayers. Her brothers Frank, Robert, and James enlisted in the Confederacy in 1861, followed by her brothers Samuel and William in 1862.

Frank was hospitalized on May 30, 1864 at Richmond and died there of disease on June 4, 1864. Samuel, wounded at Haw's Shop, Virginia on June 3, 1864, died of his wounds in a Richmond hospital on June 16, 1864. While I haven't found a war record for her brother, William, he disappears from written records before 1870 and died prior to January 25, 1877, when Tabitha's father wrote his will.

Tabitha's father, Alexander, willed the family plantation to his wife. At Rebecca's death, Tabitha inherited the plantation after paying two hundred dollars to her brother, Thomas. Alexander made mention of sons Robert and James, to whom he bequeathed five dollars each in addition to what they had already received from the estate. From that wording and the knowledge that James had already gone west and Robert had moved off the family land, we assume their father helped them purchase land or divided his land holdings at an earlier date. Tabitha was blessed to have a father who made an effort to meet his family's fiscal needs during and after of his own life. Alexander made certain his widowed daughter, Tabitha, and her family had a home after his death.

Tabitha's brother, Robert, survived the war then worked as a coach maker, farmer, miller, and door maker in communities close to his parents and sister. He and his wife, Mary, raised a family of nine.

James, Tabitha's youngest brother, survived the war then moved west. James first moved to Illinois, where he and his family were living in 1870, and then he moved to Missouri in about 1875. James and his North

Carolina wife had twin daughters born in 1869 and five other children. I doubt Tabitha ever saw James again after he left North Carolina and probably never had an opportunity to meet her nieces and nephews.

Tabitha's greatest cause for concern during the last year of the war was her husband's safety. Sanders enlisted in the Confederacy in March 1864. A gunshot wound to his shoulder put him in a Danville, Virginia hospital on May 18, 1864. A newspaper account published in Tabitha's county listed Sanders among the wounded, so we know Tabitha learned of his wound and hospitalization. If Tabitha, with two brothers dead and her husband wounded, thought circumstances could not get worse, she was wrong.

The Union Army captured Sanders on December 10, 1864, at Armstrong's Mill, Virginia. Unfortunately, prisoner exchange practiced during the early years of the war had ceased by the end of 1863. On December 11, 1864, the Union Army confined Sanders at Point Lookout, Maryland, a horrific prison where disease and privation reigned. Tabitha spent Christmas wondering about Sanders's location and situation or, if she knew of his capture, worrying about his treatment in prison. He remained imprisoned at Point Lookout until after the close of the war. To gain his release, he took the Oath of Allegiance to the Union on June 19, 1865.

Once released, Sanders had to get home. The journey back to his North Carolina family would not be easy. During the aftermath of the war, thousands of soldiers and refugees from both sides were struggling to reach their homes. Perhaps if the Yankee troops had not destroyed pertinent train rails, Sanders was able to travel at least part of the way home by train since the trains of the North Carolina Railroad had served nearby Greensboro since 1856. Somehow, he made it home to his family.

The man who came home to Tabitha was apparently ill and likely not fully recovered from his earlier wound. When I learned of Sanders imprisonment at Point Lookout, I had studied the fates of other relatives who died there. I knew immediately what horrors Sanders endured. Accounts of Point Lookout, a camp with only tents for shelter, clearly indicate six months there left men malnourished and weak or seriously ill. Many men became mere skeletons of their former selves.

Sanders attempted to resume his duties on the farm. Perhaps too weak to be careful, an injury sustained from the kick of a horse confined Sanders to bed. He died on August 7, 1865. Family lore states he died of pneumonia or tuberculosis. As Tabitha's story further unfolds, it seems more probably Sanders had unknowingly brought tuberculosis home to his unsuspecting family.

From Tabitha's viewpoint, her husband had gone to war, was wounded, imprisoned, then finally returned from the war only to sustain a farm injury and die at the age of forty. Losing her husband (in addition to the loss of two brothers who had died in Richmond the previous summer) must have been overwhelming for Tabitha. She was suddenly a single parent as were millions of other post-war women. She had five children under the age of fifteen depending upon her. There was no time to grieve.

It is apparent from the census records that Tabitha had supportive parents. Her father's first wife had died, as had her mother's first husband. From their own experience, they understood Tabitha's situation and were able to help. Empty nesters in 1860, the 1870 census indicates Alexander (seventy-seven) and Rebecca (seventy-three) had combined their household with Tabitha and their five grandchildren.

In 1880, Alexander and Rebecca (by then eighty-seven and eighty-three) were still living nearby. The 1880 census lists their household adjacent to the farm of Tabitha and her children. Perhaps what had begun as hands-on assistance in the first decade had wisely transitioned to separate but close living arrangements. In their eighties, Alexander and Rebecca knew their lives were drawing to a close, and encouraged Tabitha to stand on her own while remaining close enough to offer support and advice.

Tabitha was two weeks shy of her thirty-sixth birthday when Sanders died. She was still a young woman. Tabitha's parents had remarried after their first spouses died, then together they raised a large family. Tabitha, however, remained single and appeared to dedicate her time and energy to her children and later to their families.

Nearly fifteen years after her husband passed away, Tabitha's second child and eldest daughter, Emma, died on April 17, 1880. Emma was, perhaps, another casualty of the disease her father brought home from the war. Emma was not last of Tabitha's children to predecease her. On a happier note, Tabitha's two children who lived long enough to marry and have families of their own added a son-in-law and daughter-in-law to Tabitha's family circle. Annie Belle married Bob in December 1885. Her son, Robert, married Donna Virginia in December 1886. Because Annie Belle lost her first baby at birth, the first grandchild Tabitha welcomed was Donna and Robert's son, George, born in October 1887.

Emma Lena's arrival as a healthy baby on Christmas Day 1887 must have been as great a relief for Tabitha as for Annie Belle, who had lost her first baby at birth. Perhaps it pleased Tabitha to have her deceased daughter Emma's name kept alive in the next generation of her family. Sorrow followed joy when Tabitha's son, Franklin, died sixteen days later on January 10, 1888. If he had an illness such as tuberculosis, perhaps his death was not a shock. The official cause of his death is unknown. Tabitha's son, Cicero, died in 1896. Her daughter, Annie Belle, died of TB in 1897. Tabitha's family of five children was been reduced to one. Only her youngest son, Robert, remained alive.

You read details of the children born to Tabitha's daughter, Annie Belle, in the *Lena* and *Annie Belle* chapters of this book. The 1900 census record indicates Tabitha was living in the home of Bob (Annie Belle's widower and Tabitha's son-in-law) helping care for her four grandchildren. Two of those four children eventually died of TB as well.

Tabitha was alive when her granddaughter, Lena, married Vernon in 1905 then lost her husband to diphtheria in 1906. Perhaps Tabitha offered encouragement during Lena's pregnant widowhood and helped welcome her first great-grandchild, Robert Vernon. Her eldest grandson, George, married in November 1909. By 1910, Tabitha, eighty years old, had joined the household of her son and daughter-in-law, Robert and Donna Virginia, and seven of their children. Robert lived past his ninety-fourth birthday so he must have inherited his mother's healthy genes.

Tabitha's great-grandchildren continued to arrive. Grandson George and his wife, Donna Florence, experienced the family legacy of bitter mingled with sweet. Their son, Robert, was born in 1910 and died two days later. Tabitha lived to see their second son, a healthy boy named James, born in January 1912. James was the last new family member Tabitha met. I hope Tabitha was present when the widowed granddaughter she'd raised, Lena, married again in October 1911.

Tabitha died on March 2, 1912 at the age of eighty-one. She had lost her husband when she was only thirty-five years old then endured many more challenges and difficulties. She's a woman whose life story proves the adage that hard work won't kill you. Tabitha's life illustrates the positive impact of Godly women crossing generational boundaries to help rear grandchildren. As a young widow, she received her parents' help, and then she helped her widower son-in-law after Annie Belle died. She became a mother figure in my grandmother's life. When Lena taught me to cook, I was inheriting lessons from Tabitha.

In Tabitha's last home with her son Robert and his family, her advanced age limited her ability to perform physical labor, so perhaps she taught lessons of diligence and charity. During my research, I met two male descendants of Robert and Donna. George and Conrad are positive, life-loving people who have contributed much to their communities and families.

Tabitha's formal Christian training occurred in Methodist churches with perhaps a bit of Presbyterian worship in her earliest years. She worshiped as a Methodist with her parents, her husband, children, and her grandchildren. She was the person who took my grandmother, Lena, to Methodist camp meetings that became Lena's annual spiritual retreat. I don't have direct knowledge of Tabitha's personal spiritual habits, yet as my Granny Lena's mother-figure, perhaps Lena's habit of daily Bible study came from Tabitha's teaching or example (directly or taught to Annie Belle who then taught Lena). Surely, Tabitha needed God-given fortitude to survive the challenges she faced.

Because Tabitha is the earliest woman in her line whose detailed life we'll explore, her strong ancestral women make an appearance here. Tabitha's mother, Rebecca, was the third of five children born in North Carolina to her parents, Thomas and Tabitha (we'll call her *Tabitha Sr.).* Before we visit the life of Rebecca, we'll investigate Tabitha Sr., a woman her namesake knew well.

Tabitha Sr. was born in 1775. During her childhood, she worshiped in an Anglican church then became Methodist. She married Thomas in about 1790 and had five children who grew to adulthood. Her husband died in 1850 leaving her with a farm to run. The 1850 census lists Tabitha Sr. living alone at the age of seventy-five. In 1860, she was eighty-five years old, still living alone, had real estate valued at $2,000, and had $10,290 in personal property. She was a wealthy woman in 1860. She died later in 1860 and was buried next to her husband in the Bethany Methodist Church (established 1834) cemetery near Staley next to the church where I assume they worshiped. My great, great-grandmother, Tabitha, may have inherited more than a name from her maternal grandmother. Tabitha was thirty-one years old when her grandmother, Tabitha Sr., died. Perhaps Tabitha Sr. helped her namesake granddaughter grow into the strong woman she became.

In October 1815, Tabitha Sr.'s daughter, Rebecca, married Alfred, whose ancestors were prominent men with professions such as county sheriff, lawyer, and minister. Like Rebecca's paternal ancestors, Alfred's family had migrated to North Carolina from Maryland. By 1822, Alfred was dead.

In November 1822, Rebecca married her second husband, a widower named Alexander. The union of Rebecca and Alexander blended strong family lines from Scotland, Ireland, and England who arrived in North Carolina via Pennsylvania, Maryland, and Virginia. The society their families created in the Piedmont region of North Carolina exemplified their love of religion, democracy and education. Rebecca and her mother, Tabitha Sr., descended from women who bravely crossed an ocean, settled in a new country, moved on to better land, created homes, and managed to pass on DNA that reaches all the way to me.

Tabitha is unique among the eight great, great-grandmothers whose stories appear in this book. We're connected in what is called our *female line* because it is uninterrupted by male ancestors. Grace, Lena, and Annie Belle join me to Tabitha. Tabitha connects me to Rebecca, Tabitha Sr., and earlier women. The women of that female line have come alive to me through my research.

Markers bearing those women's names as far back as Tabitha Sr. dot North Carolina cemeteries next to places where Christians still worship, even if newer buildings have replaced those known to my ancestors. I know Tabitha instilled a love of attending Methodist services and camp meetings in her granddaughter, Lena. When Tabitha's family members suffered and died around her, she had the strength of her relationship with God and the support of her family and church. How else could she have endured? Perhaps Tabitha walked a camp meeting sawdust trail to receive the free gift of salvation through Jesus Christ. I'll want to hear the whole story when we're together for eternity.

If you're reading these words and you're a woman who finds herself helping raise grandchildren, perhaps they are too young to thank you. As a granddaughter blessed by my grandmother, Lena, who received nurture and training from her grandmother, Tabitha, I thank you on behalf of grandchildren you're blessing today. Spiritually strong women can change the world for good even if some generations are unable to do their part. Tell one grandchild of God's plan of salvation today and help future generations inherit your faith.

Mariah

1815 – 1862

When Mariah was born in 1815, North Carolinians were returning to earlier spiritual roots. Revivals and camp meetings were turning people away from alcohol and gambling to recapture the more church-going, moral society the original settlers set out to create in the mid-1700s. As the story of Mariah's life unfolds, those lofty values become apparent among her family members. Evidence indicates her parents, siblings, and extended family were Christians who were diligent in the pursuit of their vocations, and wise in their selection of spouses from equally strong religious and moral backgrounds.

Named Sarah Mariah at birth, her first name honored her father's mother. The name Mariah doesn't appear to have family origins and must have been difficult for her family and friends to spell. Mariah's call name frequently appears as Moriah or Maria. We'll refer to her as Mariah, the name spoken to me by her granddaughter, Lena.

Mariah was born on April 13, 1815. The first two winters of Mariah's life, 1816 and 1817, were among the coldest years of Guilford County, North Carolina's early history. Mariah's mother, Elizabeth, who had given birth to three children before Mariah, gave birth to a baby during each of those cold years and had another baby in June 1819. Perhaps having children in such rapid succession during years of severe weather caused Elizabeth's health to deteriorate. On December 16, 1819, Mariah's mother died.

Levin, Mariah's father, faced Christmas 1819 with seven children and no wife. Levin's mother died when Levin was in his early teens. His father, Reuben, was living nearby and perhaps could offer comfort and advice but not much more. Reuben had remarried after his wife died. By 1819 when Levin became a widower, his father and stepmother had ten more children and the youngest was under age two.

Mariah, motherless at the age of four, might not have remembered her mother. At the time of their mother's death, her siblings were Sallie (thirteen), Betsy (eight), James (six), Martha (three), Matilda (two), and Levin Jr. (six months). When their mother, Elizabeth, died, maternal grandparents weren't available to help either. Their maternal grandfather passed away at the age of thirty-one when Elizabeth was only sixteen months old. Had he lived, he might have been a good influence. His will mentioned his Bible and other books. He also included a poignant message asking his heirs to use some of his money to purchase headstones for the graves of his parents and brother at a nearby Presbyterian church. His heirs obeyed his wishes. I've visited that cemetery and seen the stones his funds purchased.

In his will, Elizabeth's father allowed his wife, Martha, the use of his plantation during her lifetime; however, Elizabeth and her brother were to inherit the land. Because Martha was only twenty-seven and expected to remarry, Elizabeth's father wanted to make certain the land remained in his family. Court records indicate Martha unsuccessfully attempted to contest the wording of her husband's will. It is unclear whether Mariah knew her maternal grandmother because Martha's date of death hasn't been determined.

Mariah's maternal uncle, James, remained close to his niece after Elizabeth's death. Many years later, there was evidence Mariah's children and stepchildren knew Uncle James well. When James passed away in 1845, Mariah's husband mentioned his death in a letter to one of Mariah's stepdaughters then living in Missouri.

Mariah's father remarried on November 5, 1823. Her father's second wife, Margaret, gave birth to Mariah's three half-siblings: William in

1825, Addison in 1827, and Hannah Jane in 1829. Family dynamics in Mariah's childhood home and in the home of her father's youth, made the blending of siblings and half-siblings seem normal to her. God was preparing Mariah for the challenges of her adult life.

What was Mariah's religious upbringing? Her ancestors appear to have begun their North Carolina spiritual lives as Scot Presbyterians. Methodist evangelism and church planting of the early 1800s touched their lives because by Mariah's time, they were Methodist.

On August 28, 1838, Mariah (then twenty-three) married Frederick, the wealthiest man in the county. Frederick, forty-seven, was twice her age. Two months before Mariah was born in 1815, Frederick married his first wife, Sarah. Mariah spent her childhood as a neighbor of Frederick, his first wife, and their nine children. Frederick and Mariah's marriage record provides a clue about Mariah's acceptance by her husband's family. A brother of Frederick's first wife served as bondsman. The participation of Frederick's brother-in-law indicates the marriage had the approval and blessing of his widow's family.

Who was Frederick? Born in 1791, he was seven years old when his father died. His mother, unable to raise him, bound him out to a farmer. Frederick was a hard worker who learned skills and diligence from the wise Quaker farmer. He served in the War of 1812, then returned to the farmer's family and married the farmer's daughter, Sarah, on February 27, 1815. Sarah and Frederick welcomed nine children between 1815 and 1836. One son died within a year of his birth. Sarah died on December 27, 1837, leaving Frederick with eight children ranging in age from twenty-three to two.

Frederick apparently wanted a wife to help raise his younger children and perhaps wanted more children. He didn't wait quite a year. Upon her marriage to Frederick, Mariah had an instant family of eight children. A month after Mariah arrived, Frederick's daughter, Louisa, married on September 29, 1838, at the age of fifteen. The next month, Frederick's eldest daughter, Mary (twenty-three and only a few months younger than Mariah), married and left home. Meanwhile, Mariah was busy giving birth to her own children. Ten months after she married, Mariah gave birth to Elizabeth Jane on June 23, 1839. On September 17, 1839, Frederick's youngest daughter by his first marriage died. Mariah's first baby was almost three months old when her stepdaughter died. If disease were the cause, Mariah must have feared for the health of her newborn.

By 1840, one son and one daughter of Frederick's first family had passed away and two daughters were married. Frederick's other five children by his first wife remained in Frederick and Mariah's home until they married. Listed in their birth order and noting the year they married, they were John in 1844, Rachel in 1851, William in 1854, Sarah in 1861, and Henry in 1860.

Mariah's life became a succession of births and deaths. Her first son, Thomas, was born on April 28, 1841 and died before 1850. Another daughter, Mary Sophronia, was born on March 17, 1843. On March 18, 1844, Mariah's father, Levin, died. She had another daughter, Margaret Lovella, on April 18, 1845. When I see Margaret's name, I remember a line from Frederick's letter to Margaret's older half-sister, Louisa.

Louisa, who married in 1838, had moved to Missouri in 1843. Frederick, in his letter to Louisa penned on June 18, 1845, wrote, "We have one more daughter in the family, about a month old and we have not named it yet." The baby was two months old (not one month old) on the day he was writing the letter. We'll forgive Frederick. I suppose that when, like Frederick, your child count has reached a dozen, you become nonchalant about specific dates of birth. Why the two-month delay in naming the baby? I know people delayed the naming process because so many babies died in infancy. Perhaps little Thomas's early death made Mariah and Frederick wary. Even though this was not the first instance of delay in naming a baby I'd seen, I had a new thought. What did they call the little darling before she had a name? Did they call her *the baby*? Surely, Mariah did not call her baby *it*, as Frederick had written.

On May 7, 1847, Roswell Jordan was born followed by Eunice Caroline, affectionately called Callie, on May 23, 1848. Giving birth to six children in ten years must have been exhausting. Mariah took a bit of a break. On October 25, 1853, my great-grandfather, Robert "Bob" Frederick, was born. Bob joined the family during the same year his father opened a gold mine in their community. Perhaps Frederick needed more money to support his ever-growing family. Mariah had one more child, Calvin, born October 5, 1855.

Mariah died at the age of forty-seven on May 23, 1862. A Methodist minister, Rev. Joseph Causey, officiated at Mariah's funeral. I'd love to know what he said. Perhaps he recapped the obituary you'll read at the end of Mariah's chapter. Rev. Causey surely preached to full pews on the day of Mariah's funeral. Her myriad immediate and extended family members alone could fill the church, leaving few seats for friends and fellow church members. Frederick, having fathered seventeen children, lived another dozen years then died on September 9, 1874 at the age of eighty-three.

Having quickly recounted the events of Mariah's complex family life, I'll slow down to provide details about the environment of her married life. Frederick, well established in his community when he married Mariah, continued to prosper. A self-made man who rose from poverty to wealth, he was apparently a creative, diligent worker. In addition to the gold mine I mentioned earlier, he owned a copper mine. He shipped copper ore extracted in North Carolina to England for smelting. A skilled woodworker, Frederick created fine furniture and coffins.

His real estate holdings grew to nearly twelve square miles of land. Frederick named the home where he raised his family Goshen Farm. His extensive pastures held horses, cattle, and hogs; he grew wheat, oats, and corn and peaches. In a March 5, 1845, letter to Louisa (the daughter in Missouri), his remarks indicated he was happy the winter had been mild, because the peach trees were already blossoming. During his lifetime, Frederick's land became the center of a rural township that was, and still is, known by his surname. The post office address was later renamed Pleasant Garden.

Frederick specialized in cured meat. He purchased as many as 120 hogs at a time that were fattened, butchered, hung in large smokehouses he built for that purpose, then sold in his store. In one year alone, his operation produced sixteen thousand pounds of cured meats. He was a country merchant who traveled to Philadelphia to purchase goods for sale in his North Carolina community. Frederick raised cotton and encouraged his neighbors to do the same. He owned the first cotton gin built in his section of the county. He was also principal owner of the banking business in nearby Greensboro, until the American Civil War years forced bank closures. Frederick also served as a magistrate and as justice of the peace. Frederick's businesses created many jobs, and his employees had the benefit of eating the acclaimed cured meets as well as the crops they helped produce.

Wives of busy husbands know Frederick's businesses and community activities kept him away from home a great deal. Meanwhile, Mariah's days were filled with work she accomplished herself as well as supervising servants in completing domestic tasks such as cooking, baking, sewing, cleaning and laundry. Gathering groceries for meals must have been easy for Mariah whose pantry was her husband's adjacent store. Women of her era provided religious education for their children and servants. She nursed the sick members of her household. Women typically supervised the feeding and care of livestock as well as the processing of their byproducts, such as churning butter or making cheese. Seldom idle, she probably quilted or completed other needlework projects in the evenings by candle or oil lamp light. How do I know Mariah was not idle? Her diligent, productive husband admired her.

Mariah's industrious husband was also sensitive and cared deeply about personal relationships. Frederick's letters to his children revealed a man who asked questions about where they were living and how well they liked their new homes. He wrote about his buying trips, and provided information about his crops, yet he never made comments about himself. Rather, he conveyed a great deal of information about the wellness (or illness) of people the recipients of his letters knew and loved. He provided information about marriages, births, and deaths. He wrote about friends and relations who had moved to new locations. He seemed genuinely concerned about writing a meaningful letter; when he felt he had run out of subject matter, he concluded with an explanation that he had nothing more to write that was worth the reader's attention.

In a stroke of good fortune a few years ago, I met a kind woman named Mary Ann who had been researching the history of Mariah and Frederick's hometown, including Frederick's role as a leader and businessman. Mary Ann had acquired a photograph of Frederick, and graciously provided a copy to me. There was an immediate feeling of connection when I saw characteristics in Frederick's face that resembled those of his granddaughter (my grandmother, Lena), whose face I knew well. Mariah's face remains a mystery.

What were the political highlights of Mariah's lifetime? She was born when our fourth president, James Madison, was in office. Her father voted in four presidential elections prior to Mariah's marriage, though her mother and stepmother had no direct voice in government. After her 1838 marriage, Mariah might have discussed politics with her husband without sharing his privilege of voting. Abraham Lincoln was in office when Mariah died in 1862. Based on the family sentiments of loyalty to the Union, her husband had probably helped Lincoln win in 1860. To honor Mariah and all female ancestors not permitted to vote, women who have that privilege must vote without fail.

What was happening in the spiritual lives of Mariah and her kin? Family stories spoken by people who descend from Mariah include accounts of attending Methodist church homecomings and camp meetings. My grandmother described those gatherings as spiritually enlightening events followed by fellowship with her extended family. Final resting places for Mariah's parents, siblings, and some of her children are Methodist cemeteries next to the churches where they worshiped. As I've mentioned, Mariah's ancestors appear to have had Presbyterian roots. When she married Frederick, he was probably Quaker, as was his first wife's family. I haven't been able to confirm whether he had been Quaker prior to his marriage into a Quaker family. His first wife's burial took place where her parents' graves were located, in the cemetery of Centre Friends Meeting.

I assume Frederick worshiped in the Methodist church with Mariah from the time of their marriage. Clearly, according to her obituary, Mariah had been worshiping in a Methodist church since childhood and there was no mention of her participation in any other denomination. An 1865 letter written by Mariah's oldest daughter, Lizzie Jane, reveals clues about the spiritual lives of Mariah's children. Lizzie Jane was writing to her half-sister, Louisa, then living in Missouri. Mariah had passed away only three years earlier so perhaps Mariah's influence colors the words and activities of her children.

To understand the letter excerpts, one must first imagine the writer's 1865 frame of reference. During the American Civil War, difficulty with mail delivery meant people might not hear from loved ones for several years. Writers were attempting to make certain family members knew about significant news items even if they were several years old. In a letter to Louisa penned on August 15, 1865, Lizzie wrote, "my dear mother died the 12th of May 1862." Lizzie wasn't aware Louisa had learned of Mariah's death two years earlier.

Lizzie wrote of the family's spiritual condition, beginning with her own effort to serve the Lord for the previous five years. Since Mariah was alive until May 1862, those five years included the last two years of Mariah's life. Lizzie reported that all her sisters had made professions of faith except the youngest, Callie, who was seventeen years old at the time. Lizzie went on to write that her brother, Milton, had made a profession of faith but their father and other brothers had not. Lizzie expressed her desire (and perhaps Mariah's deathbed wish) that their family be complete in heaven.

Lizzie wrote she was thankful to God that none of their brothers had lost their lives in the war, although she was sad Louisa's son had died. Lizzie went on to express her opinion that the war was cruel and should never have happened. Lizzie informed Louisa that the war diminished Frederick's hard-earned wealth and that troops plagued their family when headquartering on Frederick's land. She wrote that people in their area had struggled to have enough provisions during the war years, and that clothing was still scarce.

Frederick was sad and lonely without Mariah; however, he might have been glad his dear wife didn't suffer the worst of the war years. He had worked tirelessly to provide for his family and believed he had accumulated a sizeable inheritance for his children. His correspondence indicates he worked to rebuild and recover as much as possible in his later years.

Bits of family spiritual history also appeared in letters exchanged among family members in the 1880s. Mariah's stepdaughters and daughters reported the names of friends and relatives who attended Methodist camp meetings, noting names of people who made professions of faith. They sometimes included critiques of the sermons and personalities of various visiting preachers. Their interest in worship, and in the spiritual condition of their friends and family members, probably began during Mariah's early guidance and became her life-long legacy to them.

Frederick's grave is next to Mariah's in Pleasant Garden Methodist cemetery, where graves of her parents and many other family members are located. I assume Frederick, at the end of his life, was worshiping with his children who had chosen to become Methodists or perhaps to honor Mariah's wishes. It gives me pause, however, to think of Mariah's stepchildren who might have been hurt that their father did not request his burial next to their mother in the Quaker cemetery. With Frederick's two wives buried in different cemeteries, either choice for Frederick's resting place could have left some children feeling sad.

Mariah's children and stepchildren chose husbands and wives of faith. Some of Mariah's descendants joined other excellent protestant denominations that were church homes of their spouses. Whatever form of worship they chose, Mariah's descendants appear to have faith at the center of their hearts and lives. Mariah's daughter, Mary Sophronia, married a well-known Methodist minister who performed the wedding of Mariah's son, Bob, and his wife, Annie Belle. One of Mariah's daughters and one of her stepdaughters married into a family of people who descended from a circuit-riding Methodist preacher.

Nancy, a dear North Carolina cousin, provided Mariah's obituary. I'm so grateful to have this lengthy, insightful view of Mariah's life as seen through the eyes of those who knew and loved her. Following are excerpts from that obituary:

> *She leaves a devoted husband and children to mourn their irreparable loss. Many relatives and a wide circle of friends will cherish her memory, as a noble incentive to virtue and piety. Gifted by nature with a quiet unobtrusive purity and elevation of temperament and intellectual worth which, unconscious in her, gave her a controlling influence in society. And this influence was directed and consecrated by genuine piety.*
>
> *In early youth, she experienced a change of heart and united with the Methodist Protestant Church, of which she continued to be a worthy member until she was called to join the church triumphant in heaven. She was a Christian lady. As a daughter, a wife, a mother, a friend, a neighbor, she was not only without reproach; but she left an example, which in the circle of its influence, will be perpetuated to the end of time.*
>
> *The messenger who called her to heaven came in the form of measles, followed by pneumonia. For several days preceding her release, her*

> *sufferings were very great; but she submitted in patience, knowing in whom she trusted. She is not dead but sleepeth; she sleeps in Jesus; she has entered into rest. That throbbing heart is stilled in the quietings of the grave; the soul which animated it is peaceful and glorious in the presence of God. The example and influence of her whole life beckon her husband, children, friends, and neighbors to meet her in heaven.*

Mariah's good character and her spiritual strength were apparent during her life and, thankfully, preserved in writing at her death. Mariah probably never knew the Egyptian meaning of her name: God is my teacher. What an excellent name. After reading Mariah's obituary, it seems those who knew her believed she had lived as though God were indeed her teacher. Mariah's life was crowded with complex human relationships. She must have taken the most advanced classes to deal lovingly with everyone in her life.

God is still teaching. He invites us to listen and learn.

Miriam

1820 - 1877

Miriam's name, which means *bitterness* in Hebrew, has no identified family history. I assume it honors the Biblical sister of Moses and Aaron. Unlike the Biblical Miriam, thought to be the eldest child, my great, great-grandmother was the last child born in a family of ten. Between my name and Miriam's name on my pedigree chart are the names Grace, Marvin, and John. Miriam and her husband, Andrew, are the parents of my great-grandfather, John. Earlier chapters about Julia (John's wife), Lena (Marvin's wife), and my mother, Grace, introduced some of the people you'll meet again in Miriam's story. Miriam is the last woman in her line to have a dedicated chapter in this book so we'll also investigate a few of her ancestors.

Miriam was born on December 19, 1820 in Guilford County, North Carolina where she spent her life. Her father's parents were members of the Religious Society of Friends (Quaker) who knew their Bible characters. Biblical names, like that of her father, Solomon, appear among their children. Her mother, Tamar, also descended from Quakers who did more than attend their local Quaker meeting. Several of Miriam's maternal ancestors left impressive footprints of faith.

Miriam's maternal grandmother, Rachel (Tamar's mother), was the daughter of Margaret Esther, who was born in October 1721 in Pennsylvania.

When researching Margaret Esther, I saw the phrase "recorded minister" associated with her name in Quaker minutes. Unfamiliar with Quaker practices and terminology, I asked a Quaker library assistant to explain.

From the time of the early days of the Religious Society of Friends, the volunteer told me, they believed God (not men) appointed ministers and the people simply *recorded* that fact. Ministry was open to men and women equally with no laity/clergy separation. By recording Margaret Esther as a minister, her meeting indicates acknowledgement of her consistent communication of Biblical truths.

Did Margaret Esther receive family encouragement for her spirituality? One impressive character in her family was her first cousin, John, born nine months before her, who embarked on a spiritual journey that led him to write essays of faith and a journal documenting his growth and thoughts. Margaret Esther's New Jersey-born cousin died of smallpox in his early fifties during a spiritual visit to England in 1772. His writings were immediately revered and publication began as early as 1774. The values he expressed are spiritually and timelessly relevant. Republished, they form the basis of contemporary college studies of early Quaker teachings as well as spiritual principals in general. The story of Cousin John was one Margaret Esther knew well and repeated to later generations.

Margaret Esther also had a younger brother named William who had been active in Quaker ministry since his youth. William traveled as a minister in America. William sailed to England to make a religious visit in 1771 and had stayed for the same 1772 Quaker annual meeting that Cousin John had traveled to attend. William died of smallpox in September 1772 in England followed by the death of Cousin John in October 1772. John lived barely long enough to make note of William's death in his journal.

Margaret Esther lived until 1796 so she had many years to appreciate the spiritual leadership of her kinsmen and to develop her own spiritual niche. Other interesting people and events were part of the Quaker meeting where Margaret Esther worshiped. On May 28, 1768, Miriam's grandparents, Rachel (a daughter of Margaret Esther) and James were married at New Garden Meeting in Guilford County, North Carolina. Quaker people are excellent record-keepers who faithfully record many events in the lives of the members of their meetings including marriages, births, and deaths. Eight days before Rachel and James spoke their wedding vows, another event took place among the members of New Garden Meeting. Little Dolley Payne was born on May 20, 1768. Although she and her parents moved to Virginia during her infancy, North Carolina claims Dolley as a daughter of the Tar Heel State. She later became first lady of our Nation and you know her as Dolley Madison.

Rachel's daughter, Tamar (later Miriam's mother), was born in 1777 and had nineteen years to know and appreciate the spiritual wisdom of her grandmother, Margaret Esther. Tamar was present in Quaker meetings to hear Margaret Esther offer spiritually inspired words. Miriam only knew her godly great-grandmother, Margaret Esther, through her mother's voiced memories. Miriam's children received a double dose of Quaker heritage when she married into another strong Quaker line. Before we move on to her husband's family, let's spend a little time thinking about Miriam's life as she was growing up.

Miriam was the tenth and last child in her family. Tamar had given birth to Miriam's three brothers and six sisters beginning in 1798. As far as I have been able to discern, Miriam's family produced healthy individuals most of whom lived to be elderly. During Miriam's lifetime, her siblings parented more than forty children, many of whom lived in Miriam's neighborhood.

Let's pause to see what was happening with Miriam's older siblings to help frame her childhood family relationships. Introduced from the eldest to the youngest, we'll look at each sibling's life then return to Miriam and her married life.

When Miriam was eighteen months old, her eldest brother, Jonathan, married in 1822 and remained in Guilford County, North Carolina, living near Miriam until his death in 1872. An active Quaker, Jonathan served as librarian for his meeting in the 1850s. Rachel, Miriam's eldest sister, married in 1831, had at least one child, and remained in North Carolina well past Miriam's lifetime. Rachel died three days past her 85th birthday in 1885.

Born in 1802, Miriam's sister, Hannah, married in 1823 and gave birth to eight children in North Carolina. Hannah and her husband, Robert, moved their family to Indiana in 1841 where she gave birth to the last two additions to her family. Her children remained in Indiana or moved on to Nebraska. Her husband spent his career as a gunsmith and farmer. They educated their children, many of whom (and their descendants) became teachers or entered professions such as law and banking. Miriam probably didn't see Hannah after her move to Indiana in 1841 although she received first-hand reports of Hannah's Indiana life. In 1863, Miriam's son, William, went to school in Indiana where his teacher was his first cousin, John, one of Hannah's children.

Esther, was born in 1804, married in 1823, had eight children, and remained in Guilford County where Miriam lived. She died in 1861 when Miriam was 41 years old. Rebecca, child number five in Miriam's sibling group, must have been enticed by her sister, Hannah, to go west. Rebecca married in 1827, had two children in North Carolina, and then by June 1842 she and her husband were in Indiana, where they added at least two more children to their family. By the time of her death in June 1878, Rebecca and her husband, William, were living in Kansas. It is unlikely Rebecca returned to North Carolina after moving west.

Miriam's brother, Asa, married in 1836 and remained in North Carolina. He and his wife, Elizabeth, had at least eight children, one of whom they named Miriam. We know Asa left the Religious Society of Friends because by the time of his death in 1895, he was Methodist. Miriam's next brother, named *Dicks* to honor an ancestral surname, married Nancy in 1837 and remained in North Carolina. They had at least three children. Dicks' grave marker at Centre Friends Meeting bears the familiar inscription: "Weep not – He is not dead but sleepeth."

Elizabeth, Miriam's sister born in 1816, married in 1832, had at least three children, and lived in Guilford County, North Carolina until her death in 1905. Margaret, born in 1818, married Lindsay in 1841, had at least four children, and lived until 1905. Margaret's grave, along with graves of several of her siblings, is in the cemetery at Centre Friends Meeting in the still-lovely North Carolina countryside. When Margaret married in 1841, Miriam became the only child left at home with her parents.

What parental environment did Solomon and Tamar provide? Family members held extensive libraries so I'm certain education was a priority for Miriam and her siblings. Being the youngest in a large family probably meant the older siblings helped teach Miriam and she could have attended

Quaker schools such as New Garden Boarding School, a co-educational school opened by the time she was seventeen. Solomon and Tamar followed their ancestors into active Quaker membership and brought up their children to be people of faith.

At the time of Miriam's marriage to a man named Andrew on September 4, 1845, something went wrong. Members of her local Quaker meeting decided her marriage was not proper and in January 1846 the minutes indicate the meeting disowned Miriam for *mou*. That abbreviation means *married out of unity* indicating members of her Quaker meeting were expressing their disapproval of her marriage. Perhaps it was a contested decision since it appears to have taken four months after her marriage for the meeting to record the comment. Andrew's Quaker background makes their meeting's disapproval of their marriage doubly puzzling. Some reports indicate the issue was Andrew's Cherokee ancestry.

The disapproval of his marriage reportedly upset Andrew greatly. I believe Miriam and Andrew eventually found a way to reconcile with their meeting and continued worshiping at Centre Friends Meeting. One piece of evidence is, however, puzzling. A November 1873 notation in the Quaker meeting minutes indicates they received Miriam into membership during that month, but there's no mention of Andrew. Had Miriam been gone from Centre membership since 1846? Names of family members appear in various records (such as witnesses to weddings) between 1846 and 1873. If Miriam did not leave in 1846, when did she leave in order to be returning in 1873? There's no mention of Andrew, who was alive at the time.

Andrew's parents were members of New Garden Meeting in Guilford County, North Carolina, when they died, and fellow Quakers raised their children. Joshua and Abigail, a childless couple who were members of New Garden Meeting, raised Andrew who was six year old when his parents died. When Andrew grew up and chose Miriam to be his wife, he was marrying his foster mother's kinswoman. Abigail was a second cousin of Miriam's mother, Tamar. Miriam and Andrew appear to be a good match; however, the leaders of their Quaker meeting must have known something my research failed to reveal.

A deeper look into Andrew's background revealed his mother's Cherokee ancestor married a Quaker leader. Andrew's maternal grandmother, Anna, was the daughter of Hardiman, a well-known Quaker man, and his wife, a Cherokee woman called Phoebe, reportedly a member of a North Carolina sub-tribe of the Eastern Band of Cherokee. Hardiman, outspoken about his faith, was highly esteemed in Quaker circles. Some accounts refer to him as a Quaker missionary who expressed his spiritual wisdom in writing. Hardiman's documents, considered significant, are preserved in a university library.

Andrew's life prior to his marriage appears exemplary by Quaker standards. He grew up in the home of foster parents, considered pillars of their Quaker meeting. Joshua and Abigail raised Andrew with Isaac, an orphaned nephew. A well-documented activity of the two boys during their teen years was their participation as workers in the Underground Railroad. Believing teenagers were less vulnerable to searches on the road, the community asked young Andrew and Isaac to drive a false-bottomed wagon they had constructed. Andrew and Isaac made several trips taking formerly enslaved people (who traversed dangerous areas hidden in a compartment under a wagonload of cornmeal and other goods) from slavery in North Carolina to freedom in Ohio.

In an interesting discussion with Kay Davis Coltrane, author of the wonderful book *Centre Friends, The Legacy of the Meeting on the Hill* (hereafter called *Kay's Centre book*), Kay mentioned another theory about Andrew and Miriam's issues with their meeting. Andrew, according to some cousins, felt inferior because of his Cherokee ancestry, and therefore thought himself unworthy to be active in Quaker circles. Absent documentation, there is a general belief Andrew did attend worship and we know he did sign as a witness at Quaker weddings. To the best of my knowledge, Miriam's parents approved of Andrew as a son-in-law.

Andrew helped organize early Republicans in North Carolina, was a supporter of the Union, and wanted government to stay out of his personal life. Obviously, outside Quaker circles, the 1850s and 1860s found him at odds with his many of neighbors philosophically and politically; a stand that I imagine played a role in Miriam's life and the lives of their children. Even though North Carolina struggled to remain in the Union, eventually the state did secede and join the Confederacy. Many Quakers fled west while Andrew and Miriam remained in their beloved Old North State.

Miriam, if she had time, might have been interested in her husband's political activity. She was busy having babies, keeping house, and training children. Her first child, William (named, we believe, for Andrew's grandfather), was born in November 1846. In March 1848 she gave birth to Mary Fatima, called Fatima. If Miriam and Andrew were honoring Mary, the mother of Jesus, their choice of the child's first name is quite understandable. Fatima seems an exotic-sounding name for a North Carolina girl of 1848.

In 1848, the most widely known woman named *Fatima* appears to have been Fatima, daughter of Mohammed, founder and prophet of Islam. Revered as pure, some consider Mohammed's daughter the matriarch of all holy men descending from Mohammed. Far from knowledgeable on the subject, I'll attempt no further explanation. Perhaps Miriam and Andrew studied world religions. By joining the names Fatima and Mary, they named their first daughter for women greatly revered by two of the world's major religions. Fatima never married, worked as a postmistress and taught school while keeping house for her father and siblings after Miriam's death, and died in 1890.

In March 1849 my great-grandfather, John (named for Andrew's father), was born. He had one son, Marvin (my grandfather), and four daughters. Detailed accounts of the lives of John's children are included in the chapter about John's wife, Julia. In *Kay's Centre book*, she included two items that provide insight into John's relationship with Miriam.

This first item illustrates John's knowledge of scripture he may have learned at Miriam's knee. John said a farmer and fellow Quaker had given him a scriptural measure (described in Luke 6:38) when John purchased seed oats. John complimented the man's honesty by saying his basket, when full, was packed down and more oats were added.

The second story illustrates John's love of musical instruments. You will recall from Julia's chapter that John had a tendency to bring home musical instruments that Julia felt compelled to put away on the Sabbath. Apparently, Miriam had the same problem with John when he was a youngster. During the American Civil War, when John was a teenager, he found a drum dropped by Confederate soldiers when they fled ahead of the Union army. John, to Miriam's disapproval, brought the drum home to entertain his family and neighbors. Miriam, disapproving of the military sound of the drum, sent word to a Yankee commander that she had a drum for him. The last line of the story in *Kay's Centre book* was, "John had to give up the drum." John later said his mother's "Quaker gentleness" caused him to lose his drum.

Joshua (named for Andrew's foster father) joined the family in March 1851. In March 1853, Miriam gave birth to Tamar Abigail, named to honor two revered women, Miriam's mother and Andrew's foster mother. Sadly, little Tamar Abigail lived for only three months and eight days. In 1855, Miriam gave birth to Martha (perhaps named for the Biblical character). Martha married well, lived nearby, and started her family during Miriam's lifetime. In 1858, Miriam gave birth to another son, Eugene Caldwell. Although the name Eugene appears to have no family significance, Caldwell was Andrew's middle name.

In 1861, Margaret Dora was born. Called Dora, perhaps her first name honored Margaret Esther, about whom you read earlier. A church record indicates Dora's close spiritual relationship with Miriam. In December 1878 (a few months after Miriam died), Dora requested membership in the Quaker meeting Miriam attended at the time of her death. I don't know whether Miriam voiced a specific request or if Dora wanted to honor Miriam by mirroring her faith. Dora married Edgar, a distant cousin in her mother's family, and had at least four children. Dora died at the age of fifty-two when her youngest child was fourteen.

Shubal "Shoe" Edgar, born in October 1863, was only thirteen when Miriam died. He married three times and fathered fourteen children by two of his wives before dying at age sixty-four. His descendants are legion.

The family, regardless of their political posture, did not escape the ravages of war. By the time the war broke out, Miriam had seven children in her home. While Miriam was pregnant with Shubal, her eldest child, William, left home in the summer of 1863 to attend school in Henry County, Indiana. William didn't see his baby brother until Shubal was at least two years old.

William did not notify his parents he had enlisted to fight in the war. Perhaps Miriam's friends and relatives living in Indiana attempted (by sporadic wartime mail) to notify her of William's enlistment. Two years without a word from her eldest son must have left Miriam worried and prayerful. Thankfully, a letter from William survives and we know what Miriam learned in 1865.

In a letter to his father, penned near Pulaski, Tennessee, on June 10, 1865, William wrote that after arriving in Indiana, he attended a commercial school where John (a son of Miriam's sister, Hannah) taught, and worked in a drug store. He had gone to Indiana to attend school while avoiding conscription into the Confederacy as well. He wrote that in December 1863, he enlisted in the Indiana Volunteers and had fought for the Union.

Mothers whose sons have gone to war must know how helpless and worried Miriam felt if she knew he had enlisted. Her firstborn had gone to war and the Confederate home guard forced her husband into service. Among his duties, Andrew (on horseback) carried messages about Union troop movements to the Confederates. While Andrew helped others communicate, William's 1865 letter indicates Miriam and Andrew had received no direct communication from him since he left home in the summer of 1863. As William suggests in his letter, his parents must have assumed he did not survive.

William's letter is a lengthy, graphic recounting of his war years; including his capture and imprisonment. His Confederate captors took his watch, money, and pictures of his parents then confined him to the Confederate prison at Andersonville, Georgia. William reported 11,000 of 31,000 men imprisoned with him died of starvation. Stories of the horrifying conditions in Andersonville prison are readily available in histories of the period. William survived Andersonville and the Confederates moved him to Florence, South Carolina, then to Charleston. When finally released, William was so ill and malnourished he couldn't walk straight.

William's letter stated he was returning to Henry County, Indiana, and his parents could write to him there. Imagine how he longed to hear from his family. At some point before August 1870, William returned to his parents' home in North Carolina. The 1870 census names William among the family members living in his parents' household but he had no occupation. He had come home alive but ill with tuberculosis that claimed his life in December 1871. Perhaps he had never regained his health after Andersonville.

When I realized Miriam died at the age of fifty-six in 1877, I assumed she died of tuberculosis as well. North Carolina cousins who have heard oral histories through their parents and grandparents agree. I assume Miriam nursed her desperately ill son upon his return from the war. Having her son return home where he could live his last days surrounded by his loving family must have been preferable to learning he died in prison; however, his return with deadly TB robbed the other children of years they might have otherwise enjoyed with their mother, and made a premature widower of her husband.

After thinking of Miriam enduring wartime and the death of two children, let's turn to happier events in Miriam's life. Her mother, Tamar, lived until 1850 so she was alive when Miriam had her first three children. Her mother's presence and advice was important to Miriam since, as the youngest child, she had no opportunity to watch her mother care for babies. After Miriam's mother died, her father lived for another decade. Solomon knew seven of Miriam's nine children before he died in September 1860. She was pregnant with her eighth child when her father passed away.

Family lore indicates Miriam taught her children to read and memorize the Bible. They studied Shakespeare in addition to basic reading, writing, and arithmetic. Unlike modern mothers who choose to home school their children during the day, Miriam's only time to teach her children was at the end of her own busy housekeeping day, and after the children's chores on the farm were completed. Fortunately, Quaker schools offered additional education for her children. Perhaps Miriam taught her daughters they could choose a life of independence. She and Andrew appear to have provided the higher education necessary if their daughters chose a career rather than marriage. Their eldest daughter, Fatima, served as postmistress of their township in the late 1860s, and then by 1880 she was teaching school.

Miriam's daughter, Martha, married a neighbor named Frank in 1871. On October 16, 1872, Martha gave birth to her first child and Miriam's first grandchild, Leland. In 1874, Martha gave birth to Elmer then in 1877, Edwin was born. Martha's family grew to include ten children but the rest were born after Miriam's death. Martha died in 1940 and her grave is near Miriam's grave at Centre Friends Meeting.

Miriam's fourth child, her son Joshua, married six weeks after Miriam's death, so perhaps Miriam knew his wife, Roella. Ten years after Miriam's death, Joshua (Republican) served in the North Carolina State Senate. He was a farmer, a member of the State Board of Agriculture, and helped create educational institutions with a focus on improving education for African-American people. Described as tall, reserved, and solemn but full of wit, he made friends easily. In later life, he was a railway Mail Clerk for thirty years. Joshua and Roella were active Quakers who supported their meeting in a tangible way by donating land.

Joshua and Roella had two children, Roscoe and Grace. Roscoe married Ethel, a spiritually strong woman and mother of their three accomplished sons. One son became a well-known news correspondent whose name, even decades after his death, signifies excellence in his field. Miriam and Andrew would be even more pleased to learn that same great-grandson is remembered for his support of equal rights for African-American people, his opposition to the witch hunts for supposed Communist sympathizers, and his efforts to help Jewish people escape persecution and death by helping them flee Europe prior to and during World War II.

Another of Joshua's grandsons was a highly decorated man of the military who served his country with distinction. Joshua's third grandson was a successful businessman whose well-known leadership abilities led to his appointment to manage vital activities in his home state and recognition by his governor. All Miriam's great-grandchildren who descend from her son, Joshua, appear to have become assets to their churches and communities as have the generations who follow them.

I mentioned the birth of Miriam's eighth child, Dora, who married a cousin from Miriam's family. Dora's husband, Edgar, was her third cousin, twice removed. That union produced four children; only two of whom lived to marry and have children.

Miriam's son, John, married eight months after Miriam died. Although she wasn't present for the marriage ceremony, Miriam knew John's bride, Julia, whose story you've read.

When Miriam died on July 11, 1877, two of her nine children had already passed away, one had married, and six remained in her home. Andrew, left to finish raising the younger children, must have depended on their older daughter, Fatima for help. Her dedication to help raise her siblings explains Fatima's spinsterhood. At the time of Miriam's death, Eugene was nineteen, Dora was sixteen, and Shubal was thirteen.

Thirteen years after Miriam's death, two of her single adult children died in a four-week period. As mentioned in the story of his sister-in-law, Julia, Eugene, who had been working as an Internal Revenue agent, died of wounds received in a raid on a distillery. A February 1890 newspaper article about his work reported Eugene and another agent destroyed ten illicit distilleries in a neighboring county. In the earlier raid, moonshiners reportedly fought the agents and fired about thirty shots although no one was injured. On June 24, 1890, Eugene (then thirty-two) died of wounds and injuries sustained when he fell from his horse during a raid in Stokes County. On July 20, 1890, his forty-two-year-old sister, Fatima, died. A July 31, 1890 newspaper article reported that Fatima died of grief over the death of her brother.

Would Miriam's presence have made a difference to Fatima? Perhaps Andrew, grieved at the death of Eugene, was stoic and missed the signs that Fatima was overwrought. Of course, there was surely an underlying medical reason for Fatima's sudden death. Perhaps a pre-existing condition worsened by stress, lack of sleep, or failure to eat properly caused her death. Fatima was an educated career woman by the time of her death, so she must have known people who could have offered solace. Had Miriam failed to teach Fatima to deal appropriately with grief and to ask God for peace?

Andrew survived until 1903 and apparently opened his home to his adult children and grandchildren in his later years. The 1900 census record for Andrew's household as includes Shubal and two of his sons (Shubal's wife had died of childbed fever). Andrew must have felt personal loss after Miriam's death and would have been concerned about the absence of a mother and grandmother for their children and grandchildren. Since there was no adult female in Andrew's 1900 home, Shubal's older brother and sister-in-law, Joshua and Roella, adopted his infant son. Shubal, with his father's help, raised the two older children until he married again. Shubal married a second time but his wife soon died without issue. Shubal and his third wife, Mary, produced eleven more children.

Shubal has far too many descendants for all their stories to appear here so I've selected three people to highlight. One man apparently inherited Andrew's Republican leanings. Shubal's great-grandson, Calvin Coolidge, was (unsurprisingly from that name) a Republican elected to represent his district in the North Carolina State Senate. While serving, he was also a member of the Chamber of Commerce of High Point, was co-owner and secretary-treasurer of three businesses, and (this would thrill Miriam) he was active in his church (Oak Hill Friends Meeting) where he taught Sunday school, was Sunday school superintendent, and served as clerk of the meeting. He also served as president of North Carolina Quaker Men.

Another of Shubal's great-grandsons chose a career both Miriam and Andrew would applaud. Mitchell served as pastor of Methodist churches for forty-one years. After his seminary training and pastoring in New Jersey for a short time, Mitchell pastored churches in Western North Carolina. Officially retired by 2011, Mitchell preaches occasionally for special events. Imagine Miriam greeting people on her side of heaven who can tell her they heard the Gospel from her great, great-grandson.

The third of Shubal's great-grandchildren I've chosen to appear here is Cousin Shirley. She has tirelessly answered questions about her branch of the family. Shirley has inherited, corrected, and preserved the family research of Cousin Joyce of Pennsylvania, who documented their branch of the family then passed away.

Absent details about Miriam's personality and activities outside her home, we'll look at Andrew's interests and activities. In addition to being a farmer, merchant, a Quaker, and a political activist, he spent time supporting worthy causes. In 1849, when Miriam and Andrew had been married three years, Andrew attended a County Temperance Convention. After hearing speeches in favor of the abolition of intoxicating drinks in their county, Andrew accepted appointment to the Committee on Vigilance tasked to discourage the use of alcohol in their area.

Miriam's husband was not a man to stand idly by. He got involved and took a stand when issues arose. Absent information to the contrary, I assume Miriam was in supportive agreement with Andrew's efforts. As you've read, two of Miriam's sons worked as revenue agents in 1890 to abolish illegal distilleries. Oddly, however, two decades after Miriam's death, perhaps Andrew changed his mind with respect to alcohol. His son, Shubal, who was living with Andrew in 1900, reported an occupation of *storekeeper, distillery.*

As I've learned the stories of Andrew and Miriam's descendants standing up for people who are denied rights or who have been singled out as "different" and therefore treated unkindly, I've thought of the beginning of Miriam and Andrew's marriage when they were made to feel unwelcome (perhaps because of a few drops of Cherokee blood). What an excellent example of God's grace. Without personally experiencing discrimination, they might not have become people with backbones of refined steel. They produced descendants willing to stand firmly for what is right.

Clearly, God had a plan when he combined the strengths of Andrew and Miriam to generate positively productive offspring. Strangers have written stories about this couple's descendants changing the world for good. Miriam's personal faith, evident when she rejoined her congregation several years before her death, illustrates our need to forgive and move on.

When I step back and attempt to view Miriam's life as a whole, I realize I envision Miriam through facts surrounding her husband and descendants without specific stories or insights about Miriam, the woman. While her husband enjoyed public pursuits, Miriam had the important role of preparing her children for life. Miriam's lessons on character and faith, passed down across generations, positively influence descendants now playing key roles in their churches and communities.

Miriam must have been breathlessly busy. She had to create clothing for her family, cook three huge meals a day from scratch, and make household items such as soap. She also gave birth nine times then, without electricity or indoor plumbing, performed all the physical tasks involved in mothering: nursing, diapering, bathing, and caring for the children when they were ill.

Miriam, while not a leading character in family stories, worked hard to accomplish the overwhelming tasks she faced every day. She did not leave a personal letter or spiritual journal yet Miriam is important to me as my kinswoman. When we give our best effort, our contributions, even if not remembered as peerless, are of vital importance. God has a plan and our individual parts don't always reach the public stage. We might not hear thundering applause until eternity.

Speaking of eternity, after Miriam shuffled off to those streets of gold, Andrew never remarried. He died on the last day of 1903, over a quarter century after Miriam's death. Although Andrew appears to have attended events held by his Quaker meeting, records are silent on the subject of their meeting's initial disapproval of Andrew and Miriam's marriage. Andrew's remains rest next to Miriam's in the graveyard of pretty Centre Friends Meeting. I look forward to seeing them worship together in eternity.

Even if Miriam imagined her descendants living upright lives and achieving their goals, she might be awed at the multi-generational recognition of a great-grandson who spoke his mind to help ensure individual freedoms. She might be even more impressed to learn two of her great, great-grandsons are helping men find spiritual salvation then leading them to physical salvation as they overcome drug and alcohol addition. She might not understand the medium, but she'd be amazed to learn her great, great-grandchildren share prayer requests by e-mail. Miriam has a role in everything her descendants accomplish. Without Miriam stepping into her place, filling her role, and living her life, what might the world have missed?

Quiet, modest, and leaving almost no personal history, she passed on her character to those who came after her. Unlike the biblical Miriam whose disobedience meant she missed her earthly reward, I trust our Miriam is now in the heavenly Promised Land hearing "well done, good and faithful servant."

Cordelia

1838 – 1881

My search for the story of this great, great-grandmother was initially so frustratingly fruitless, that I was tempted to give up. I confess I did put the failed task aside several times. But I'm no quitter. One morning, I took a deep breath, and decided to take a fresh look at the lives of Cordelia's daughter, Mamie, and Cordelia's other children in search of clues.

I decided to begin with Mamie's younger sister, Susie, who had become an educator. Since a high school in Accomack County, Virginia, bore Susie's name, I hoped the school's history might include Susie's biographical information (including the names of her parents). My brief e-mail to a county school administrator generated results far beyond my expectations.

The recipient of my inquiry forwarded the note to a man affiliated with the history department of Susie's high school. Based on his research of families in their area, he knew of my family, and was acquainted with other living descendants. He telephoned me and said, "I can do better than provide information about Susie. I can introduce you to your cousins." He then provided the name and phone number of my cousin, Patsy.

With my heart pounding, I dialed the telephone number, and soon heard the sweet voice and lilting Virginia accent of Cousin Patsy. I felt right at home. Patsy and her daughter, Nancy, are faithful guardians of our family history. In addition to her own first-hand knowledge of the family oral history, Patsy inherited (and was willing to share) hand-written letters, photographs, and scores of other ancestral artifacts. Thanks to the efforts of Patsy, Nancy, and the dear women who came before them (especially Emma, Cordelia's youngest daughter); I've learned far more about Cordelia and the people who populated her life than I had dreamed possible. This chapter would be brief and dull without the generosity of these gracious southern cousins.

Patsy and Nancy welcomed me by phone, communicated via e-mail, and then hosted my visit to their area. They interrupted their busy lives to drive me to ancestral land grants and remote family graveyards I would never have found on my own. Patsy allowed me to spend several days and nights gleaning treasured stories from her wonderful family history files, and then recited unwritten stories while I took notes.

Cousin Patsy made certain I met Cousin George, a descendant of Cordelia's oldest daughter, Bessie. George, a leader in business and government, shares our interest in family history. He hosted a cousins' luncheon during my Virginia visit, and provided wonderfully detailed information about his branch of the family. Patsy also connected me with Mary Thom, the dear wife of Cousin Winter who descends from Cordelia's brother. Mary Thom graciously provided documented results of her research of Cordelia's parents and ancestors.

Among the first bits of enlightenment I received from Cousin Patsy were the correct call names of the individuals who inhabit this branch of our family tree. She had the benefit of living among family members who spoke of our ancestors by name. Cordelia, called *Cordie,* also referred to herself by that shortened form of her name in several of her letters (yes, *Cordie's letters,* written as early as the 1860s). You'll read some of Cordie's letters in this chapter. First, we'll learn about Cordie's ancestral families and learn about her life prior to her letters.

I've included insights about Cordie's mother-in-law, Elizabeth, and Elizabeth's ancestral roots. The ancestors of Cordie's father-in-law, Nathaniel, were included in the story of Nathaniel's granddaughter, Mamie. Because Elizabeth died three months after Mamie was born, Mamie didn't know her paternal grandmother. During Cordie's childhood, she knew Elizabeth as a neighbor, and then the two women had a close relationship during the first eleven years of Cordie's marriage to Elizabeth's son, Tom. With those introductory remarks in place, let's begin at the beginning of Cordie's life and look back at her roots.

Cordelia Elizabeth was born on April 15, 1838, on her father's ancestral farm, Willowdale, in Accomack County, Virginia, on the Delmarva Peninsula. Researched and co-authored by Mary Thom, a beautifully detailed report documents and preserves the history of Willowdale. Mary Thom, in addition to researching her husband's branch of our family, participated in a project that resulted in the designation of Willowdale as a National Bicentennial Farm in 1991. The Bicentennial Farm award commemorates farms remaining in the same family for more than two hundred years.

According to Mary Thom's research, when Cordie's paternal ancestor, John, arrived in America in 1665, he purchased 500 acres of land. Cordie's father, Hugh, inherited the land and his ancestral home in 1837. Beginning with an initial structure in 1728, successive generations expanded or modified Willowdale's structure to fit their needs, and to maintain a livable dwelling for their large families. Although long vacant and neglected, a structure still stood when I began researching Cordie's family and first visited Willowdale. Across the grassy field from the family home, I walked among the gravestones of Cordie and other members of her family. Long-dead ancestors appear to keep watch over the home they loved. Since my visit, a distance cousin had begun renovations and now occupies the estate.

Who were Cordie's parents? Her father was Hugh, son of George and Elizabeth. George was a well-known man who, in addition to farming his land, served as a justice in his county in his later years. Hugh's paternal grandfather, John, fought in the American Revolution. Hugh's mother descended, on her mother's side, from the Custis family of Virginia (sharing ancestors with the first husband of Martha Dandridge, whose second husband was George Washington).

Having found little information about Cordie's father, Hugh, I read a lengthy obituary written about Cordie's brother, Frank, who was the next heir to the family land. If Frank's manhood reflected the character of his father, Hugh was well-loved, generous man of integrity. An American Civil War story about Hugh indicates his neighbors held him in such high regard they risked their own lives to rescue him from Yankee captivity.

Cordie's mother, Margaret, descended from families who came to America early in the colonization period. Margaret's father descended from a couple who arrived in Jamestown in the early 1600s. A son of the

Jamestown couple moved to the Eastern Shore of Virginia, acquired land in Accomack County, and produced several generations of heirs including Margaret's father, Richard, who died when she was three. Margaret's mother, Esther, descended from immigrants who settled in Virginia in the mid-1600s. Esther's parents passed away before 1813 when Esther, then seventeen, married Richard. Esther, widowed at twenty-three, married again. Her second husband died five years later and, still in her twenties, Esther married a third time. Esther's third husband was dead before 1850. After losing three husbands, Esther managed her own land and household until at least 1860. Esther was living with her daughter, Margaret, at the time of her death in 1870 at the age of seventy-six.

Cordie's parents, Hugh and Margaret, married in 1832. Their first child, Susan, died at the age of four on March 31, 1837 and their second child, James, died at the age of two on April 1, 1837. Two children dying on consecutive days may indicate an infectious disease. Margaret and Hugh persevered and soon Margaret was pregnant again. Cordie assumed the position of first and eldest child when she was born in April 1838. Her siblings and their years of birth were Frank (1841), George (1843), Oran (1845), Peggy (1848), Susan (1851), Harwood (1855), and Tom (1857). Oran died at the age of two; George died at twenty. Cordie's other siblings attained adulthood.

Cordie's first foods included locally grown figs, sweet potatoes, corn, and peas. Other local crops during her childhood were rye, wheat, tobacco, and cotton. In the 1840s, nearby towns of Belle Haven and Pungoteague grew to include churches, mercantile stores, and physicians. Tradesmen such as tanners, shoemakers, and blacksmiths opened businesses. Conversation around the dinner table might have included topics such as construction of nearby lighthouses, gristmills, lumber mills, and a brick-making plant.

By the time Cordie was thirteen, schools in her county numbered twenty-seven and were educating over twelve hundred students annually. Cordie and her siblings attended a school at nearby Coal Kiln, according to Mary Thom. Schools offered reading, writing, arithmetic, and geography classes to individuals between the ages of six and twenty-one. Cordie's parents also hired private tutors to supplement their children's public education. Cordie, as evidenced by a self-portrait that hangs in Cousin Patsy's home, learned to draw and paint.

Cordie appeared with her parents in census records only twice. She appears as a child under five in 1840 (before first names of children were included), then she appears named among her parents and siblings in 1850. On March 8, 1857, she married a neighbor gentleman, Littleton Thomas, called Tom. When a woman marries, she gains a husband and his family. Cordie's mother-in-law, Elizabeth, descended from early Eastern Shore Virginia families. Thanks to Elizabeth's sons who wrote detailed descriptions of their parents, we know the physical attributes of Cordie's mother-in-law and a bit about her personality.

Elizabeth, described as a woman of medium height with a good figure, had black hair, greenish blue eyes, slightly angular features, a fine nose and a firm, quiet mouth. Although not physically strong, she lived past middle age by the force of an indomitable will combined with a cheerful disposition. She was friendly and self-possessed with a dignified yet kind, gentle manner. A fearless woman of deepest sympathy, she was unimpressed by the favor of others, and had fine judgment, tact, and discretion.

Elizabeth was the boss of the house and she was fond of smoking a pipe. She wore hand-knitted, knee length stockings and kept her pipe (when not in use) inserted in the top of her stocking with the bowl protruding from the top. When friends came to call, it was her custom to pull up her black silk skirt, pull out the pipe, and light up before discussing neighborhood affairs. She died of tuberculosis at the age of sixty on Wednesday, November 4, 1868.

Having explored Tom and Cordie's families, let's view their married life. The 1860 census locates Tom and Cordie's little love nest in Accomack County where they had grown up. Tom was listed as a farmer, Cordie was keeping house, and their first child, Elizabeth (noted in that census as Bettie and later called Bessie), was one year old. Not long after that census, Cordie gave birth to Edwin (called Ned). Their world was about to be rocked by war.

Tom and his brother Nat were outspoken abolitionists yet, as loyal Virginians, they enlisted in the Confederacy during the American Civil War. During her high school teaching career, Tom and Cordie's daughter, Susie, asked her father to write of his 1861 Confederate service. Tom's notes indicate Susie planned to use the account to teach her students about the history of their area. As Tom put it, he left his wife and children and obeyed the call to gather with his neighbors to fight for the Confederacy. In a lengthy account, Tom states he reported for duty, drilled, and took charge of a group of men.

According to Tom's account, the Colonel in charge of the Confederate troops decided to surrender. Their small ill-equipped, ill-trained force was doomed when faced with the well-trained, well-equipped Yankees, who outnumbered them nearly two to one. Tom, angered by the surrender, considered the Colonel's decision a show of cowardice. He wrote a poignant account of dismissing his men, shaking each man's hand, then breaking his sword and walking home. A fellow soldier picked up the sword, took it home, and later returned it to Tom.

Tom reported that during the Union occupation of the Eastern Shore, many men slipped through the naval blockade, crossed the Chesapeake Bay, and joined forces with the Confederate Army. I have found a badly spelled record of a man who might be Tom fighting with a Richmond regiment. I know Tom wrote to Cordie from Philadelphia in 1864. Perhaps he had been asked to supply the troops with food by beginning the produce business that later became his career. Let's catch up with Cordie's life then we'll get back to Tom.

After the war began in 1861, Cordie's childbearing continued. On May 19, 1862, Cordie gave birth to her third child, Hugh, named for her father. Little Hugh died in April 1863. Her fourth child, George Hugh, was born on January 10, 1864. In keeping with family traditions of the time, George Hugh's name honored Cordie's brother, George, who had passed away ten days earlier on December 31, 1863 at the age of twenty. Brother George was the last sibling Cordie lost in her lifetime.

Ten months after George Hugh was born, Tom was in Philadelphia when he wrote to Cordie, who was still in Virginia. We know several men in Tom's family eventually moved to northern cities to build a business enterprise they formed to export produce grown in their home county to other parts of the country. Although Tom doesn't explain what he was doing in the North in 1864, his letter doesn't indicate he is serving as a soldier on either side. In addition to facts conveyed in the text, Tom's letter gives us a sense of the warm, gracious tone he employed when communicating with Cordie.

Philadelphia, October 22, 1864

Dear Cordie:

I did not intend to let so much time pass before answering your letter but I have so many things to detain me from writing sometimes when I am in the midst of it I am interrupted and fear I shall be today. I received your letter on Thursday in which you enclosed Harry's (letter). *I was so much obliged to you; he sent me 6 Confederate stamps and I am going to send him some U.S.* (stamps). *He is well and in better health, more so than he has been for some time, still at Monticello.*

How I should like to see your roses now and if it were not so far and late in the winter I would bring you some. Do not apologize for your paper. I am glad to get so long a letter and I fear it would not be so long if the paper were smaller. I am glad to hear of Billie Bundick's getting religion. I expect his parents are rejoicing at it. Write soon and as often as you can and tell me all the news for am anxious to hear from my friends. Love to all. Kiss the children for me.

Respectfully I am
Your friend, LT

Cordie had apparently written about their neighbor *getting religion,* and Tom comments on that topic in his response. This is the earliest comment on faith I've found in correspondence to or from Cordie. During wartime, Tom and Cordie were concerned not only about the physical health of their friends but also about their spiritual well-being.

Life for Cordie's family must have been difficult during the years of Yankee occupation. Yankee troops requisitioned churches and schools for their headquarters and living spaces during the occupation. Yankee leaders permitted churches to hold worship on Sunday mornings provided their service included a prayer for the Union and President Lincoln.

On October 20, 1865, Cordie gave birth to a daughter, Margaret Elizabeth "Peddie," named to honor the child's grandmothers (likely present at her Virginia birth). Peddie's birth date indicates Cordie and Tom were together in early 1865, so Tom left Philadelphia not long after his October 1864 letter. Before Cordie gave birth again, the scenery changed.

You know our family has letters from the 1860s because you've read a letter written by Tom. To my great amazement, Cousin Patsy had several letters written by Cordie herself. I'll share the meat of those letters with you; however, to better understand Cordie's environment and times, let's begin with some introductory facts.

Cordie and Tom moved north at a time when northern cities were trying to repair their former alliances with southern exporters, including the produce growers Tom represented, to feed their growing populations. The letters indicate Cordie was in New York as early as July 15, 1866. Cordie went back to Virginia in late 1866 through early 1867. She again wrote from New York beginning in mid-1867 through April 1869. I've included only examples of those letters since they are similar in content.

What did Cordie see and experience in New York City in the late 1860s? Although improvement efforts were underway in older sections of New York, refuse often collected in the streets to knee height. Thousands of horses added manure and urine to already unsanitary streets. Attempting to avoid outbreaks of diseases such as cholera, New Yorkers had begun moving out of the most crowded areas. Tom secured rooms for the family in Greenwich Village, where they shared the streets and housing with freed slaves, budding artists and young writers. Among New York City accommodations of the era, Greenwich Village supposedly offered a cleaner, small-town-like atmosphere.

When their boat neared New York in 1866, Miss Liberty was not yet in the harbor to greet them. They lived near Washington Square although there was not yet a Washington Arch. New York University had been operating since 1831 near the Washington Square Park where Cordie might have taken her children for a stroll. Central Park, completed in the early 1860s, was becoming a popular spot where families could picnic. There was no Brooklyn bridge yet and no George Washington bridge.

Buildings on Wall Street might have looked tall to my rural Virginia ancestors, although photographs from the era depict buildings four or five stories high. Trinity Church, now dwarfed by skyscrapers, appeared to be soaring above other buildings during Cordie's New York years. There are New York City photographs taken as late as 1869 depicting mansions and such iron and glass structures as the then-named Grand Central Train Depot, with cows still grazing nearby.

Even though the telegraph was becoming available, the late 1860s were still clearly an era of hand-written correspondence. Daily letters between loved ones were often the norm. Although hundreds of letters were written, we're grateful for a few surviving examples that provide intimate views of Cordie's daily life.

Using family documents and census records, I've been able to identify and confirm locations of people Cordie mentions in her letters. Maggie, a frequent addressee called *sister*, is actually Cordie's sister-in-law (Tom's sister). Further, when Cordie speaks to Maggie about *Father and Mother*, she is referring to her in-laws. When she sends greetings to other members of the household, they are people in Tom's family home. Cordie's standing as a member of Tom's family is apparent at the close of her father-in-law's life; Nathaniel willed the family home to both Tom and Cordie. Other people who make frequent appearances in Cordie's letters are Tom's brother Nathaniel (Nat), and his wife, Lizzie Susan; and Tom's brother, Edwin, and his wife, Mary Henry.

Without further delay, I'll provide tidbits of family letters mingled with other happenings and information, organized chronologically, to help convey the story of Cordie's all-too-short life. I'll refrain from interrupting the flow of Cordie's letters to explain names or words, except when a bit of clarification will help convey a significant relationship or point. Enjoy reading letters about Cordie's daily life in post-war years.

New York, July 15, 1866

My Dear Maggie,

We are fixed at last. I mean done cleaning, putting down carpets, etc. I am also nearly through fitting up the orders from friends at home. We have bought no furniture except a round walnut table $5.50, and a suit of cottage furniture for $36 and a stove $24. The rooms are very comfortable, and I feel very well satisfied.

How are you all getting along? Do write often and tell me all the news. This fall, we hope to have fine times. Give my love to Mary Henry, and tell her I have been too busy to get her music, but Lizzie Susan and I are going early in the morning. I hope to be able to send all by Capt. Carter. The ink is in use, and I am using pencil in preference to waiting.

The church bell is ringing, and Tom is sleeping soundly. The children all seem to like it here very much. Lizzie Susan is anxious to hear from Nat. Tell Father and Mother to be sure to come this fall. Give my love to them and to all.

Your sister,
Cordie

When Cordie penned the next 1866 letter, Tom was in Virginia and she was awaiting his return to New York. It must have been comforting that her sister-in-law lived nearby; the two women weren't alone when their husbands traveled to conduct their produce exchange business. Everyone's health seems to be a constant consideration with Cordie. Her letters often ask about the health of the recipient as well as bearing a report on the health of members of her own household.

We'll move on to 1867 when Cordie provides specific insights about her spiritual life; she includes comments about the pastor whose teaching she enjoyed.

New York, May 26, 1867

My Dear Maggie:

I have not forgotten my promise to you, although I did not write last week. I have had a very busy time shopping, fixing our things, washing, etc.

We went to hear Dr. Deems last Sunday morning, and to hear his brother in the evening, Dr. Deems' lecture on husbands and wives last Wednesday evening, and to hear him preach again this morning; so you see we "stick to him" as Tom says. If you hear him once, you want to hear him again.

They sing old-fashioned hymns and wind up with the never failing "collection" but they stand very much in need of help, as they have to hire the Chapel of the University. It is a splendid building. If we were able, we would have a pew.

Has the Legett (ship) *arrived? I am afraid Edwin's fruit was all rotten. I put in a pineapple, half a dozen oranges, one can of peaches, one of tomatoes, one tongue, and three pounds of crackers (two of cream and one of wine crackers). Now if he wants more of any or all, I want you to let me know. How glad I would be to hear you all were coming up and Edwin with you. I hope he will be able to come before the summer is over.*

Give my love to all. Nat's family all went on board the Duncan yesterday to go down (to Virginia). *The wind is ahead and raining. I fear they will have a tedious trip.*

When you write, tell me if Miss Ada has ever got her bonnet and if it suits her. Write as often as convenient, Maggie. You do not know how I love to hear from home. When I write again, I shall tell you something about our ridiculous fashions.

Your sister,
C. E.

The history of the church Tom and Cordie were attending is quite interesting. They were hearing Dr. Charles Deems, formerly a North Carolina Methodist minister, who organized The Church of the Strangers. Members professed the fundamental truths of Christianity without adhering to the restrictions of sects or denominations. The church was perhaps one of the earliest non-denominational churches in our country.

On July 22, 1866, sixteen people (including the pastor's family) met in the small chapel of New York University on Washington Square, a few blocks from where Cordie and Tom were living. Sunday school began at nine o'clock on Sunday mornings, followed by a public service at ten-thirty. By the time of Cordie's mention in her May 1867 letter, the church's attendance had grown and relocated to the large chapel at the University where they could accommodate four times more people.

The name of Tom's brother, Nat, appears on the list of the first members to take communion on January 5, 1868, the day the church was formally organized. Tom and Cordie might have missed church that day because of illness, or perhaps they were still in Virginia following a Christmas visit. Having explored Cordie's choice of church in New York, let's return to her letters. I am pleased she took time to write about the challenges of every-day life. Because my grandmother used the term, I know Cordie's description of laundry being "rough dried" referred to clothes that had been dried but not ironed.

New York, June 24, 1867

My Dear Maggie,

I was very glad indeed to receive your letter a few days ago. It was very interesting and full of good news. We are always rejoiced to hear that you are all informing in health.

Tom has written to Father and Mother and told them all the news about his business, how he stands it, etc. Mother asks him how I stand washing. First rate, tell her. I washed about 50 pieces last Friday. They were all small; however, I only wash for George and Peddie and our handkerchiefs, collars, and stockings. It generally takes me half a day but they look very nicely when finished. I think if it was winter instead of summer I should have no trouble as the heat bothers me worse than the labor. They charge 50 cents per dozen for rough dried clothes and such as I wash count two for one, but you see 4 dozen pieces would count 2 doz. which would be $1 and by doing it myself, I save that much every week and what I hire is usually from 50 to 75 cents. They furnish soap and starch and if you count the coal, it is cheaper to have it done out. It is almost impossible to get a good washwoman that will come to your house, and they charge $1.50 a day and you have to attend to their fire, make their starch, and clean up after them and having the heat to contend with. We do not bake much but when we do it is on ironing days.

I would like to hear that you are coming with Edwin and Lillie. What a nice time I shall have when you all get here. I do hope Edwin will be able to come soon. Mother wrote he would if his strength would permit. And she also said she and father were coming after you all had your trip. Now that should be an inducement for you all to hurry up.

Tell Miss Ada I have had Early York Cabbage, and thought about her when I was taking it up for it was cooked like I have heard her say she likes it: thoroughly done and greasy.

Tom is thinking of making a change in his business next year. In fact, he has fully determined to do so, but of course it is a secret yet and to that end, we are saving every penny we can. I have bought myself a bonnet and common hat. The two (cost) *$1, which is all the wearing apparel any of us have bought since we have been here. I had enough left of my black muslin for a sac and have made a very pretty suit of it. I have not been to church since strawberry season commenced for Tom has to sleep on Sundays now. It is indeed a day of rest for him and he needs it too. I wish there were two Sundays in one week. Give my love to all.*

Yours,
Cordie

I commend Cordie's wish for "two Sundays in one week." She apparently wanted a day of rest for her husband, yet longed for a day of worship. In the midst of her own taxing housework, she was concerned for her husband's physical needs during his demanding work season.

Tom's brother, Edwin, died on July 28, 1867. On August 31, 1867, Tom's sister Molly died. Married to a distant cousin in 1861, Molly and her infant son died at (or shortly after) childbirth leaving a husband and three-year-old daughter.

Names Cordie mentions and phrases in her text place the penning of an undated letter in the mid-1868 timeframe. Cordie comments on her need to rest part of the every day. I believe Cordie needed rest because she was in advanced pregnancy. She gave birth to my great-grandmother, Mamie, on August 18, 1868.

The next excerpt was part of a letter that did not survive in its entirety. Although we have no postmarked envelope and the letter is not dated, clues in the text helped me date the letter to December 1868. Cordie indicates she is writing shortly before Christmas, and describes Mary (my great-grandmother, later called Mamie) as baby. Mamie was four months old in December 1868.

They (her children) *have jolly times, but have to be kept under some restraint constantly. I hope sometime to be able to have a basement for them, where their romping will not disturb other families below them. Little Mary is a great hit with all, and I really believe has cut Ped* (Peddie) *out with her Pa. She is so good and playful that she is a general favorite.*

I have had three turkeys sent me from the (Virginia) *Shore as Christmas presents. I think including my desk and oven, I came off well. Mr. S. sends Tom the turkey because he was so well pleased with the sale of his turnips.*

Give my love to Miss Adah. Tell her I would make some mince pies if I knew how to make them like those we used to have at Atlantic View (Tom's family home). *Dear, dear old home.*

We will soon have all the family hanging around the oven and it is great satisfaction to us. It is so cold that we have had to put our bed in the front room and keep two fires. It is snowing fast now and I dread to go out. I would not mind it so much if I went oftener, but I stay indoors so much that this keen air goes hard.

Bettie is suffering with a cold but nothing serious now. The rest of us are all well. Night is coming on and I must turn nurse. Tom is jumping Mary and spoiling her, making more work for me tomorrow.

Write as often as you can. I know you have had your hands full lately.

Tell Mannie and Allie (her children's cousins) *the children often speak of them now-a-days. Toys are plenty and the stores look very fine to children. Goodnight.*

Your sister,
C. E.

A lengthy letter from Cordie to her sister-in-law, Maggie, in April 1869 indicates they would soon move to new rooms. I believe the move they were preparing to make was to 8 Bethune, the home address shown for the family in the 1869 New York City directory. The year 1869 marked the arrival of another of Tom's brothers in New York. Cordie must have been pleased to have more family members nearby.

Tom's brother, George, moved to New York in 1869 where he became a sanitary inspector for the Board of Health. A medical doctor who had graduated from the University of Maryland in 1864, George had initially practiced in Virginia. He apparently didn't enjoy New York life, and returned to Virginia by the time of his marriage in 1873. He later became a Virginia Senator.

After this evidence placing Cordie and Tom in New York in 1869, I've been unable to identify their place of residence in the 1870 census. Family stories place their homes from the late 1860s to 1880 in Philadelphia, Baltimore, New Jersey, and New York; interspersed with returns to Virginia. Tom's brother and business partner, Nat, was in Baltimore with his family by 1870; however, Cordie gave birth to her next child, Susan Cordelia, on September 28, 1871, in Virginia.

Tom and Cordie were in Baltimore when Eva was born on April 15, 1874, and when Emma was born on August 20, 1877. Wherever they lived, Tom was working to support his family, Cordie was raising her children, and both of them were maintaining close relationships with their extended family. Shortly after Cordie completed her own family, she lost a parent. Her father, Hugh, passed away on July 19, 1879.

As the 1870s ended, Cordie was becoming ill. Specific information about how she was feeling in the 1870s has eluded me. Some census records include notations about individuals who were ill on the day the census taker visited their home. The June 1880 census includes an entry for Tom and Cordie in Baltimore with their children, plus a niece and a young white woman who lived with them as a servant; however, there is no mention of anyone in the household being ill. A few months later, we know Cordie was gravely ill.

Thanks to Cousin Patsy, our family has Cordie's final messages to several family members. When Cordie's letters begin, I'll step out of the way and let Cordie tell the story of her difficult, final challenge. Before you read Cordie's final words, however, you'll need to know what she was enduring. Cordie had tuberculosis (TB); however, the terms *tuberculosis* and *TB* didn't come into use until 1882 when Robert Koch identified the tubercle bacillus. *Consumption* was both the name by which Cordie knew her disease and the result of the infection. TB consumes its victims. Cordie thought and spoke of herself as a *consumptive.*

When I learned about Cordie's TB, I wondered when and where her initial exposure occurred. I thought of Cordie's 1860s life in crowded New York City, and assumed her exposure to TB occurred there. Recalling that Cordie's mother-in-law died of TB in 1868, I realized Elizabeth possibly infected Cordie before she left Virginia. Once exposed, Cordie's infection could have remained latent until stress and exertion of raising a large family allowed TB to display its manifold horrors.

When Cordie became ill, she knew few consumptives survived; although many experienced symptom-free periods associated with a variety of treatments. Prevailing therapy schemes included sending patients west or south for *pure air*. Cordie had to leave most of her family behind when her husband attempted to improve her health by moving her from Baltimore to Jacksonville, Florida. Understandably, individuals and groups opening facilities catering to consumptives faced fearful citizens who fought to prevent their opening, or demanded segregation or removal of the ill from their community. We glimpse the humiliating social climate consumptives endured in the account of a confrontation with a fearful woman Cordie met en route to Jacksonville.

It was initially difficult to identify the addressees of several of Cordie's letters. No envelopes survive and salutations were often "my dear child." Using clues, I determined the addressee for each letter. By combining census findings with clues in Cordie's letters, I learned where her children lived while she and Tom were in Jacksonville. The older children, probably to attend school, remained in Baltimore with a housekeeper and the younger children were with Cordie's family in Virginia. During Christmas break from their schools, the older children were apparently going to join their siblings in Virginia.

Cordie identifies family members by call names her children knew. She mentions her brother (Uncle Tom), and her mother (Grandma), to younger children who are staying with Cordie's Virginia family. In 1880, Cordie's widowed mother, Margaret, was Cordie's children's only living grandmother.

Cordie's husband, Tom, went to Jacksonville with Cordie and they took along their daughter, Emma (sometimes called Tiny), then three years old. At first, I wondered why they took their youngest child, and then read it was because Emma was exhibiting symptoms as well. Cordie's sister, Peggy, and a servant named Maria also accompanied them to Florida.

Cordie wrote a three-part letter about the journey to Jacksonville; however, only parts two and three survive. Thanks to Cordie's graphic descriptions, we almost feel we're traveling with her. Her words paint a poignant picture of the emotional pain of illness, and the feeling of separation she had to endure. My heart ached for Cordie when I read her account of receiving cruel comments; yet Cordie remained upbeat. No matter how ill, she managed to recount cruel treatment with humor, and enjoy the beauty of the landscape around her.

Cordie and her companions traveled by boat to Charleston, South Carolina then by train to Jacksonville. The first letter fragment, Part 2, recounts what took place en route to Jacksonville. A woman on their train complained openly about traveling with an apparent consumptive. Understandably, their fellow traveler feared TB. The level of fear we sense in the woman's words paints a grim picture of Cordie's appearance.

Part 2

> (The woman was) *...nearly smothered in a gay plaid shawl. Well, the conductor then told her there was one berth in the car she had left (we needed two) and she readily jumped up to go there when lo and behold she saw Emma, then she left off I tell you. I never heard the like before and as a parting salute, she pointed her finger in my face as she passed me and said if her child died I would be the cause. While it amused the others, it was very embarrassing to me. After a while, we got off and passed through the most untouched piece of country I ever saw. The forests have bloomed out until they can't be called forest, the entire soil covered with little palms a few inches high, and one third of that is under water. On almost every little elevation, there was a log hut, such as we would not stable a horse in. As night came on, they all had fires outside (for mosquitoes, I presume).*
>
> *At one of the little depots, we saw two Negro men sleeping out of doors each on a piece of wood with a fire between them and the smoke just covering one. Your Pa went and looked at them and they were sound asleep. As we neared Jacksonville, we found beautiful residences with yards fitted with orange trees bearing their golden balls, Cape jasmine, and lilies in bloom with a profusion of roses and other flowers.*
>
> *At last, after the roughest ride we ever had, we reached our journey's end, and a fortunate thing we did not miss the train; for the next one ran off the track, and remained in that dismal place all night and nearly all the next day before relieved. I don't know how they made out for food, for the passengers only take supper expecting to breakfast here.*

Let me advise everybody to buy tickets to Savannah, and then come the inland (waterway) route which costs $1 more. The old engine would actually stop at every little elevation, the fire up, back a little and try again like some poor worn out horse. When we got to the hotel, who but the woman with that blessed baby should be next to start to enter. She jumped down as though we had been alligators, and said she would remain at the depot a week first. There were no closed carriages and as it was, we took other conveyance, which goes wherever directed. The hotel is a very pretty though plain structure. The houses are all wooden except warehouses and stores that are nearly all next to the river.

On the back of the page, Cordie wrote:

Dear Children, I mail you this and do not think it necessary to write. There is no news. We are all pleased with our new friends, but it is a little strange to be boarders in our old house. We'll write you a long letter soon.

The added note about "our old house" indicates the family had visited Jacksonville prior to the current trip. I have been unable to learn when their previous visit occurred. Tom does appear to export fruit on this trip, so perhaps he had done so in previous seasons.

Part 3

There is but little need for art for nature will do all you require in ornamentation. We have an old live oak in the yard full of long-hanging moss, its knotty roots are filled with earth, and flowers have taken possession there. It is refreshing to look at. A bright lemon-colored chrysanthemum is in full bloom with no frost to spoil it. Several stumps about the yard walk have been cut out, and flowers planted in them with vines around the sides, which not only cover the sides, but on the ground and geraniums as fresh as in June.

We were given rooms in a little cottage adjoining. They are roughly built but comfortable, and mine has the sun from early morning until late afternoon. As soon as our box came, which was delayed several days, we rented the rooms furnished and went to housekeeping. We bought a nice new little cook stove for $15. That with a small table was about all we *had to buy. We are not private. Emma has plenty of room in and out of doors and we can't complain of anything. Your Pa is getting his business arrangements fixed, and thinks he can start a profitable trade.*

Jacksonville is a perfect garden of loveliness. I hope that some day you may all see it, though in health. My throat improved so much at sea and now is so sore.

It had been raining for two weeks when we came, and that followed by a dense fog all night and until ten o'clock next morning, and I think I took cold coming into these rooms which of course were closed and no fire since last season. I hope it will soon be better. Emma's cough is better. She talks a great deal about you all, and Cousin Mannie as much as anyone. Tell her I include her in this, as I shall in all my letters, though after she leaves I shall write her separately. I am glad, very glad you are to remain together.

It is quite a task for me to write on much, and as the children in VA want to know about the route, I would like you to re-mail this. Now my dear children, take care of yourselves and try to keep well. I am not uneasy about anything else. I have full confidence in you all. Write often.

Your mother

December 6, 1880

My dear child,

I am very glad to tell you my throat has begun to improve and I am feeling better. It is a great comfort to get such interesting, encouraging letters from you all. We remember the 10th but I am so feeble that I must wait for Christmas. (Cordie's reference to "the 10th" points to her eldest daughter as the addressee of this letter. Bessie's twenty-second birthday was three days later on December 10. Cordie was too ill to shop for her daughter's birthday.)

Your Pa is writing and will tell you all the news. I have an excellent, kind-hearted physician—a cured consumptive, which led him to give special attention to that disease. His name is Dr. Fernandez and I never saw such black hair and eyes before. I want you to remember me especially to Ida, Lillie, Sudie, Sarah, Miss Phena and Lena.

*Tell Lena (*that*) Maria is delightful and wants us to come back next winter. She says Lena was to get a tie from some place and take it to you. If she gets it, put it in an envelope and a couple of stamps will bring it.*

My best love to my boys and tell them their turn comes next.

My lung trouble is in my right lung and that arm soon gives out now I am so feeble. I hope the oil, malline, etc, will soon make me stronger. In fact, I feel stronger today.

Write me any little item. All is interesting and I am so anxious to hear from Mamie. Tell all I have named to write and not wait for me—it will be charity. Did you get the house cleared and the cellar? My respects to Mrs. Scoggins. Auntie is going to write all soon. The thermometer today is at 70. Emma is better and says tell Bud and George she is good. I shall get George some orange leaves.

God bless you all.
Your mother

Jacksonville, December 12th, 1880

My dear little Eva,

I said I would begin with the youngest and that is yourself and Susie next. (Note: Eva, then age six, was the youngest child who did not go to Jacksonville.)

Well, what must I write about? You have heard all about Jacksonville, haven't you? Don't you think it looks funny to see a hog in the street? They come in sometimes, walk along, and grunt as if they felt at home. We often see men riding on the horses back instead of in the cart and it looks real funny.

Across the street is a large place, railed-in, where horses are kept for sale. Emma sits at the window often and watches them eat, play, and sometimes she claims two or three but says they are not as pretty as Uncle Tom's Max. Your Pa had to buy her a toy horse and she named it Max.

You have seen a hog killing by this time, I expect. It will soon be Christmas. Your Pa is going to send you all a box of oranges and I expect Sister (Bessie), *Bud* (Ned), *and George will send something.*

Emma says tell Susie and Eva she wants to see them. Her cough has almost entirely left. I am getting better. It has been very cold the people here think. They had a little bit of ice and it was a great talk. It is an extremely cold winter but we have no fire and a window open all the way.

Auntie is writing to Grandma. Why doesn't Aunt Sudie write to us? I want my little children to try and be as good as they can and not trouble Grandma.

You can wear your unlined dresses by putting on your under-bodies.

The first time Aunt Sudie comes, ask her please to measure your and Susie's seersucker dresses for me. I know they are much too short but I can allow. I want the length of skirt, waist, and sleeve (inside seam).

Papa sends his love to you all. I shall write Susie in a day or two.

Your devoted mother
Eva kiss Grandma and Uncle Tom for Auntie

Jacksonville, December 13, 1880

My precious son (Edwin "Ned"),

I am very nervous this morning but have put off writing so long hoping to feel better that I shall try now and do the best I can.

Letters from you children almost give us new life. I can't begin to tell you how much we enjoy them. Your Pa has just received yours of the 9th. You have heard all about the place, and there is no news for me to tell you. We rode out into the country about a mile and a half yesterday and every yard had young orange trees in it. Two thirds of them seemed to be new settlements. I am truly glad you are so comfortably fixed. I am anxious to know if you can get the cellar cleaned and those mattresses and beds, etc., carried away, i.e., as rubbish.

I think a trip to see the children at Christmas would be a very nice thing for your sister (Mamie, I believe). *Why doesn't she write? It has been more than a week since she wrote, and neither of the pictures has come. I am delighted to know the little ones are so cheerful and contented.*

Your father is packing oranges. He lost on the first and I am uneasy about all. I hope the severe weather will not make you all sick. It reached here and one morning there was a little ice in a tub, which caused a great deal of talk, the people not being used to it. It never stays cold long here; the sun has too much power.

I am improving slowly and my throat is better, but at one time was worse than it had ever been. There is a gentleman at the hotel suffering just the same way and it seemed to comfort him so much, your father said, when he told him how bad mine had been and now so much better.

*Tell George (*that*) Maria says he will have his orange leaves. The next writing I shall do shall be to him. I do hope, my dear boy, that you will get your trip. It will be a great pleasure to all. Your father likes it so well here that he wants to come every winter. Maybe we will all get settled down here. It is certainly a delightful place. Fruit retails equally as high here as in Baltimore and having a chance to buy direct from the growers, they double their money on everything. And you would hardly believe how much they use. You see someone eating them (*oranges*) from morning 'til night. A piece of orange is like a coal of fire in my throat.*

It is mail time. A kiss all around.

Your mother,
C. E.

December 18th, 1880

My darling little Susie:

It is your turn now for a letter and you will have it. I don't know what to write about because I can't get out to see new things. I know you and Eva had fine times at hog killing. You must write and tell me all about it.

Emma's cough is well now and she plays downstairs with other children.

She still pulls off her shoes and stockings and when she is too lazy to put them on herself, she says the toes keep going where the heel ought to go.

*Your Pa says if he sends you a box of oranges they will all be rotten before you can get them. He has written your (*brother, Ned) *Bud to send you a nice box of good things from Baltimore.*

I was real glad to hear through Aunt Sue's letter that you were all so well and happy. I am not one bit uneasy about any of you. Tell Uncle Tom I wish him a Merry Christmas and Happy New Year. I am afraid Grandma caught cold during hog killing. I wish we had some sausage and sauce. Auntie is going to write Grandma and Aunt Sue tomorrow.

Emma says tell you she wants to see Susie and Eva. Tell Peddie and Mamie their turn will come soon. Be good children. Give my love to the family and Aunt Sue's, Uncle Harwood's and Frank's. Your aunt Maggie too if you see her. Goodbye now, my dear little ones.

Your mother,
C.E.

*December 24th, 1880 (*the last letter we have*)*

My dear Son (Ned),

Your father shipped yesterday a box of large mangos and oranges on the Sturgis to your sister and they were marked just as the others were to Mrs. Schall and I want you to look out for it. It is a smaller box.

We are much more comfortable in our new home and I enjoy sitting out on the balcony so much. I can see that I am stronger already. I got very weak from such close confinement. I am looking forward with a heart full of joy to your coming.

The stores, they say, are handsomely fixed up for the holidays. Santa Clause and horns in many windows.

I have made no preparations whatever, not even a cake. It does not seem like Christmas one bit. I want your sister to give Mrs. Schall two large oranges and a twig if they keep. Your father could not get any moss to wet to put among the leaves. The oranges, in that case, should have been covered in sawdust.

George's letter came today, and also yours to your Father. I shall write soon. Your letters, from all, are so cheering and like sunshine after rain.

My throat is first better and then worse, hurting me right much now.

Emma has learned to eat oranges. You know she would not eat them at home. The letter is full of love to all. There is not one bit of news.

Well, too late for Christmas but Happy New Year for all.

Your mother, C. E.

On the back of Cordie's letter, Tom wrote a note to Ned on Christmas Day: "your mother suffers very much in her throat." Cordie's condition was worse than she was admitting to her children.

Cordie passed away on January 11, 1881. She was forty-two. By the time she died, TB had absorbed all the underlying fat of her face leaving her cheeks hollow and her eyes sunken. As evidenced by the clear, concise letters Cordie wrote in the weeks preceding her death, a consumption victim's mind remains quite sharp.

It must have been extremely difficult for Tom to watch his dear wife wither away after doing everything he could to seek a cure. He was always struggling to make enough money to raise and educate their children. Setting up a residence for himself, Cordie, one child, and a servant in Jacksonville and simultaneously supporting children in Baltimore and Virginia must have created additional financial stress. Tom's efforts to improve Cordie's health left him with no wife, debts, and children to materially and emotionally support.

Tom later made the following entry in the "Deaths" section of his family Bible: "In Jacksonville, Florida, amid the flowers she loved so well, (died) Tuesday, January 11th at quarter to nine o'clock a.m., my dear wife, Cordie, in the 43rd year of her life. My soul is exceedingly sorrowful even unto death."

Tom returned Cordie's remains to Virginia for burial among her ancestors at Willowdale. Cordie's grave marker bears Tom's comment that she died in Jacksonville, Florida, amid the flowers she loved. At the time of Cordie's death, she left Tom (forty-nine), and eight children: Bessie (twenty-two), Ned (twenty), George (seventeen), Peddie (fifteen), Mamie (twelve), Susie (nine), Eva (six) and Emma (three).

Margaret had lost her daughter. Because her first two children died before Cordie was born, Cordie was Margaret's first child to survive to adulthood. Margaret must have depended upon Cordie to help raise her younger siblings, and then enjoyed her companionship as an adult. Margaret outlived Cordie by sixteen years.

Among my eight great, great grandmothers, only Annie Belle, who died at thirty-nine, had a shorter lifespan than did Cordie. Yet of all the women I've researched, Cordie left more personal documents that survive. Her own words help eliminate guesswork about her personality and values. If I were listing summary thoughts about Cordie's characteristics, selflessness would top the list. Throughout her life, her letters reveal a woman who was trying to meet the needs of other people, make them feel important, and reassure them if she sensed they were sad or worried.

Once in Jacksonville, Cordie probably realized, as she compared her decline with the symptoms of people dying around her, that her life was ending. Cordie spent her remaining strength writing letters to her children. I thought Cordie insightful to tell her children she had confidence in them. Had Cordie lived to instill more of her selfless nature in her daughter, Mamie (twelve when Cordie died), Mamie might have met her life's challenges in a more outwardly positive, loving way.

Since Cordie wasn't alive to witness more of her children's lives, their later years were included in Mamie's chapter. Cordie's children remained an emotionally close sibling group who spent time together often throughout their adult lives. Her children's spouses and her grandchildren missed knowing this lovely woman.

What were Cordie's greatest strengths? She loved her family, her in-laws, people of lower social station, and even attempted to be patient with cruel people. She seemed to rise each day thinking of little deeds she could accomplish to help others. She was always thinking of the person to whom she was writing or the people around whom she was living.

She was a spiritual woman rather than a legalistically religious woman. Most of her life, she worshiped as a Methodist. She wanted to hear sound

Christian teaching and sing old (familiar) hymns, yet she didn't seek empty ritual. Amen. When I learned The Church of the Strangers appealed to Cordie and Tom, I recognized the roots of my own worship preferences. I knew what Cordie meant when she indicated she'd heard enlightened preaching and wanted to hear more.

When her husband was exhausted and sleeping through the Sunday church bells, Cordie yearned to worship but didn't awaken Tom and insist they go to church. She allowed him the Sabbath rest. Judge her decision negatively if you must; however, I suggest Cordie was making the Godly choice. Most of us have struggled and asked God's help to achieve balance. Being in church for every event could be the wrong choice if, for example, family health needs or in-depth spiritual study must take second place.

Fortunately, Cousin Patsy preserved a photograph of Cordie as a young woman. I often look into Cordie's sweet face and feel drawn to her. In my daydreams, I meet Cordie in eternity and ask her to fill in all the mystery years of her life. I want to know where she went to church in Baltimore and in Philadelphia. I'd love to know more details about Tom and Cordie's lives, and about the childhood of my great-grandmother, Mamie. Did Cordie sob into her pillow in Jacksonville when she sensed her death was near and she'd never see her children become adults, get married, or have children of their own? I have so many questions.

Before hearing Cordie's answers to my questions, I should thank her for leaving evidence of a life permeated with grace and patience. Her parents named her wisely. The Celtic meaning of Cordelia is *daughter of the sea*. What a perfect name for a girl born so close to the edge of the Atlantic Ocean. Her parents might have been Shakespeare fans who named their daughter after the character named Cordelia in King Lear. Shakespeare created his character, Cordelia, as a beautiful, honest and devoted daughter.

Cordie's combined forenames, Cordelia Elizabeth, offer a more spiritual connotation. Another interpretation of her first name begins with the Latin root of *cor* meaning *heart*. Elizabeth means *consecrated to God*. Cordelia Elizabeth means *heart consecrated to God*. Cordie's parents gave her a name that describes positive characteristics she actually developed in her life.

By her example, Cordie has taught me that our lives (no matter how short) can be exemplary. She set a high standard of gracious womanhood. What a blessing to know of her life and share her bloodline. Active love toward people is proof of our love for Jesus, and Cordie's life was a testimony of her love for others. We don't need to share Cordie's DNA, however, to read the words of Cordie's many letters and be encouraged to mirror her strength.

Tuberculosis separated Cordie from her life in this world; however, nothing separated her (nor can anything separate us) from the love of God. Before illness sapped her strength, she capably produced and trained children whose descendants continue to mirror Cordie's strengths when they love their families and neighbors, work diligently in their churches, live as productive members of their communities, and serve in government leadership roles.

Holding Cordie's original letters in my hands, I marveled that she'd never have imagined her great, great-granddaughter holding them over a century later or sharing her messages with readers of this book. Her words, penned so close to the time of her death, illustrate her graceful exit from this world. She didn't whine. While suffering, she used her last measure of strength to write loving, positive, personal messages to each of her children. She mentioned subjects she knew would make the child smile, relax, and feel the mother they loved was still with them, even though she was ill and many miles away. She departed in an attitude of joy.

Hoping her children could visit her in Jacksonville, Cordie wrote, "I look forward with a heart full of joy to your coming." I imagine Cordie saying those words from Heaven. How pleased she must be if all her children have indeed joined her in eternity, where she radiates resurrected health.

Spiritually strong women die gracefully and live eternally.

Sarah

1849 - 1907

Two sharply contrasting events occurred in Dorchester County, Maryland, in the autumn of 1849. One event was the birth of a baby girl into large, loving family. The other event was the departure of a young woman who had endured two decades of slavery and abuse.

The girl born that year was Sarah, my great, great-grandmother, the subject of this chapter. The other event marked the beginning of a heroine's mission. Harriet Tubman, a brave African-American woman, fled Dorchester County in the late autumn of 1849. Harriet was taking the first step toward becoming a conductor in the Underground Railroad. While Sarah was welcomed, Harriet was leaving behind everything and everyone she had known. She even left her husband (a free African-American man) who refused to go. Harriet was fleeing slavery and seeking freedom.

The memories these two women had of the county where they grew up, a generation apart, differed greatly. Sarah lived fifty-seven years surrounded by loving family and friends, endured little danger or deprivation, then passed into obscurity. You'd never have known Sarah's story if you were not reading it in this chapter.

If you're a student of history, you know of Harriet's determination to escape her own enslavement then free others. Once she fled, Harriet never again lived in Dorchester County, although she bravely returned several times to help transport others to freedom. She was a woman on a mission.

Researching Sarah's life a century after her death was too late for first-hand interviews. Only descendants of those who knew her were living and, while quite willing to help, had little to say. Historical accounts describe Sarah as a pleasant woman who lived and died quietly. She could have experienced challenging events that don't appear in family stories or census records; however, based strictly upon my findings (or absence of findings), she endured few difficulties. From Sarah's life-giving branch, her two children produced six branches of descendants.

We who are still alive can step out of quiet obscurity and live with purpose and bravery. If there's a worthy cause, let's fight for it. However, God doesn't call or equip every woman to save people from oppression. I believe Sarah accomplished precisely the plan God envisioned for her life with graceful humility. Sarah's life was vital to the physical and spiritual continuation of our family. Here's her story.

Sarah Emma was born on October 23, 1849. Sarah's name means *princess* and honors her father's mother. The name Emma (meaning *grandmother*) doesn't appear to have a family origin. Sarah's parents, William and Hester, spent their lives in Dorchester County, as had several generations before them.

In Sarah's era, people of the Eastern Shore of Maryland adhered to traditions of early generations. After two hundred years of living in a relatively closed society, Sarah's neighbors discussed their frequently shared family trees over quilts and piles of corn. To be safe, even without proof of relationships, they developed a habit of calling one another *cousin*.

When Sarah was born, her father was in his second marriage, and Sarah was the sixth child born to that union. If any character in Sarah's story led a challenging life, it was her father, William. Born in 1804, William first married a woman named Margaret Ann on December 11, 1828. He and his first wife had five children. During his nine-year marriage to Margaret Ann, William supported his family by farming and was active in local government. He helped arm the local militia. Beginning in 1832, he served as co-constable for his district, and then as constable beginning in 1834. Margaret Ann died on May 30, 1838, leaving William with children who were then ages eight, seven, five, three, and two.

After five months as a widower, William married Hester Ann, the younger sister of his first wife, on October 30, 1838. Among the five children William fathered with first wife was a daughter named Hester Ann. Margaret's children must known their aunt Hester well before their mother died. Since Margaret and Hester's parents had died by 1836, Hester might have been living with her sister and brother-in-law at the time of Margaret's death.

Ten months after William and Hester married, the third child of William's first marriage, Reuben, died a month before his sixth birthday. On April 28, 1840, Hester gave birth to her first child. They named the little girl Margaret Ann to honor William's first wife, Hester's sister. Margaret Ann died three months later in July 1840. Hester had more than a year to focus on raising her sister's children before she'd have another of her own.

Hester and William named their second child, born in October 1841, Reuben Henry. Unlike William and Margaret's first son by that name who died at age five, William and Hester's Reuben lived until two days prior to his eighty-fourth birthday. Ezra and Enoch (who both lived long, full lives) soon joined Reuben. William and Hester illustrated their familiarity with Old Testament Bible characters when naming their sons. Hester's fifth child was Margaret Jane, who grew to adulthood.

William and Hester's sixth next child was Sarah. The 1850 census, taken on September 16, 1850, noted William (forty-six) as head of his household with his wife, Hester (thirty-three). The youngest two children of William's first wife, Hester Ann (fifteen), and Riley (fourteen), were listed with the children of the second marriage: Reuben (nine), Ezra (seven), Enoch (five), Margaret (four), and little Sarah, who was not quite one year old on the census date. William owned real estate valued at $1,000 and was supporting his family by farming.

Ten years later, the 1860 census noted William had been busy earning money for his growing family. When the census taker interviewed the family on July 13, 1860, William reported his real estate value at $3,000 and his personal estate at $20,000. William was then fifty-five and Hester was forty-three (and eight months pregnant). In 1860, Sarah was no longer the baby of the family. Hester had given birth to John in 1853 and Henry in 1857. William was supporting a family of ten. Hester's huge workload meant Sarah, at age ten, was already performing her share of chores.

Six weeks after the 1860 census, Hester gave birth to Everett on August 29, 1860. Everett died on October 5, 1861. On May 24, 1866, Hester gave birth to another son, again named Everett, who died a few weeks later on July 11, 1866. Hester's childbearing ended. Sarah was a teenager when she observed her mother's pain at losing two children in five years. Such trauma surrounding childbearing and loss possibly influenced Sarah's later decisions about family size.

After the births and deaths of the 1860s, Sarah had four older full siblings and two younger ones in addition to four surviving half-siblings from her father's first marriage. Normally, children descending from one father and different mothers are half-siblings. In Sarah's family, the children of the two groups are half-siblings because of their common father and first cousins because their mothers were sisters. This kinship scenario creates duplicate branches on the family tree. Sarah's choice of spouse further complicates genetic branches with close, intertwining twigs.

The 1860 census lists Luther, a fourteen-year-old neighbor, living near Sarah's home. Luther, whose father had been lost at sea, was living with his widowed mother, Leah, and his sister, Emily, who was three years his senior. Fortunately, financial circumstances permitted Luther to help support his mother and sister while attending school at least part of the year. He and Sarah were probably attending the same one-room school. He's someone who plays a role in Sarah's future.

According to the 1870 census, William and Hester's household included Ezra (twenty-seven), Enoch (twenty-four), Sarah (twenty-one), and Henry (thirteen). Ezra and Enoch were farming for their father who was then sixty-five years old. William's 1870-era estate had diminished; indicating he had given money to older children when they left home to start their families, his holdings diminished in value during the American Civil War, or both. On July 22, 1870, William's real estate was worth $3,000 and he had $600 in personal funds (down from $20,000 in 1860).

William died on October 29, 1870, followed five months later by Hester who died on March 30, 1871. Sarah was twenty-one, single, and to my knowledge had no personal source of income. Her older brothers, Ezra and Enoch, may have remained at home to keep the younger children together. Fifteen months after her mother's death, Sarah had her own home.

On June 9, 1872, Sarah married her neighbor, Luther. Called Captain Luke outside the family because of his first vocation as a sailor and ship captain, Sarah may have called him Cousin Luke. During Sarah's lifetime and earlier, marriages among cousins were common in their neighborhood. Luther was a third cousin of Sarah's father so Luther and Sarah were third cousins, once removed. When Sarah married and took her husband's name, her surname did not change.

On January 9, 1873, Sarah gave birth to a daughter, Daisy. The 1880 census lists their family as consisting of Luther (thirty-four) who was working as a sailor, Sarah (thirty), and Daisy (seven). During Sarah's married life, she had help with her home. In 1880, Sarah's housekeeper was Eliza, a single woman Sarah's age. Even though financial commentary is absent from that census, we know from family records Luther had been saving to purchase his own boats and eventually acquired two schooners (named for his children).

Luther was a busy man. He operated a general store in their little town of Madison. We know Madison was then a two-store town because of a story found in family notes. According to the story, the local post office was sometimes located in Luther's store, and sometimes situated in the store of his competitor, Mr. Jones, directly across the road. Luther was a Democrat and his neighboring store proprietor was a Republican. The post office was reportedly located in Luther's store when Democrats were in office and in the other store when the Republicans were in power.

Before home mail delivery, people retrieved their mail at the post office. Particularly on Friday and Saturday, customers and neighbors congregated at the stores. General stores were among the few public places where people could meet for a chat. When people came to retrieve mail and greet neighbors, they could purchase goods Luther and the other storekeeper procured locally or imported from Norfolk or Baltimore.

Luther's stock included meats, canned goods, locally grown vegetables, clothes, shoes, hardware, candy, sewing supplies, dry goods, cakes, and pies. In addition to Luther's own boats, he used the services of a steamship, the Emma Giles, to ship goods to and from Baltimore. Upon unloading incoming goods at a nearby wharf, he transported the goods the short distance to his store by oxcart.

The location of Luther's store, next to their home, was convenient for Luther. He also departed from a nearby wharf when sailing his boats to transport pilings and lumber to Norfolk and Baltimore. On the return trips, he brought back farming tools and other merchandise to sell in his store. He employed local people to work in the store, and help with loading, unloading, and sailing his boats. Luther posted wages of his employees on his books, and then deducted purchases from their wages. The people who provided goods or services could also draw cash from their wages.

In addition to the four acres of land on which Luther built his home and store, he owned two tracts of land. Legal documents indicate Luther's tracts were originally part of larger parcels of land owned by Sarah and Luther's common ancestors. On one tract, Ed (an African-American gentleman) sharecropped. Ed also helped Luther by managing the lumber crop on the second tract, an effort that included periodic thinning by cutting out dead trees to ensure better growth. The relationship was beneficial to both men. After Luther died, Ed continued to work with Luther's children until they disposed of the property.

Sarah's level of involvement in Luther's business may have included helping him select goods she thought women would purchase. Perhaps she and her cook made the pies and cakes sold in Luther's store. I know she was raising Daisy; making certain her daughter acquired excellent schooling and domestic skills.

Raising one child from 1873 through 1889 doesn't sound daunting. Reports of Luther's career as a merchant mention no stressful issues. Sarah may have worried about her husband's safety at sea. According to family lore, Luther had many stories to tell about sailing his schooners to Baltimore, Annapolis, and Norfolk. An often-repeated story describes a harrowing trip to Norfolk with a hull full of lumber, and a deck loaded with pilings.

In Luther's era, before sophisticated weather forecasting systems, sailors began their journey having checked the usual signs familiar to sailors of the day. On one particular day, a sudden storm developed into a ranging hurricane that overtook Luther's fully loaded ship en route to Norfolk. Fortunately, he noticed the storm developing in time to swing into the wind, and get his sails furled and tied down. Under bare masts, Luther traveled the remaining distance to Norfolk in a sea of foam. At a crossing speed never matched by Luther before or after that day, he lost some of his deck load, but sustained no damage to his boat. Did Sarah see the storm approaching over the Chesapeake, and wonder if her husband would come home? Luther's father had been lost at sea, as had other family members and acquaintances. Perhaps she knew worry wouldn't help, and chose to pray for his safe return.

Sarah sometimes accompanied Luther. One family story mentions Sarah making a trip to Baltimore. Luther sold some family property, gave Sarah five thousand dollars to spend as she wished, then took her shopping in Baltimore. Sarah purchased a bag of ribbon and some feathers, and then said she wanted to save the rest of her money. What a good, frugal woman.

Luther and Sarah must have found schooling too limited in their rural location. When Daisy was a teenager, they sent her to a female academy. As mentioned in Daisy's chapter, Daisy lived in a boarding hotel near her school until her parents whisked her home (upon learning she was attracted to her much-older landlord). Sarah might have lost sleep, or at least had some serious pillow talk with her husband, over their daughter's romantic ideas.

Soon after Daisy's mistakenly placed romantic feelings, Madison Methodist Church received a new pastor named Vincent. With a home across the street from the church and plenty of food, Sarah and Luther were accustomed to providing housing and meals to their minister. They welcomed Vincent into their home. Eventually, Vincent (then thirty-three) asked for Daisy's hand in marriage and their wedding occurred on Daisy's seventeenth birthday, January 9, 1890. Luther and Sarah approved.

In the midst of preparations for Daisy's wedding and packing for her departure as a newly married woman, Sarah gave birth to the second of her two children on December 22, 1889. Eighteen days later, if she had sufficiently recovered, she was sitting in the mother-of-the-bride seat at Daisy's wedding. Sarah's only opportunity to see a child of hers wed was Daisy's wedding. Daisy's children would also be the only grandchildren Sarah lived to know. As mentioned in Marie and Daisy's stories, Sarah's home hosted the 1893 birth of Daisy's second child, Marie, as well as many lengthy summer visits by all five of Daisy's children.

As Daisy left her nest, Sarah began the mothering process anew with little Oden Luther. Oden was the name of Sarah's nephew who had passed away, and of a beloved former Maryland governor, Oden Bowie. Luther must have longed to teach his son to sail, and probably took Oden sailing as soon as Sarah allowed him go. Styles for little boys in 1890 included the sailor look; appropriate for a little boy whose father owned two schooners. Sarah could have stitched little sailor suits for Oden from fabric purchased in Baltimore or Norfolk. American Civil War uniform production had forced the development of an efficient garment-making industry, so perhaps Luther purchased ready-made outfits for Oden in Baltimore. In any event, dressing Oden as a baby was easier financially, and from a supply standpoint, than dressing baby Daisy had been seventeen years earlier.

Before leaving the subject of attire, even though Sarah lived in a rural area, visitors to the Eastern Shore in the 1870s described women in nearby Cambridge, Maryland, as fashionably attired. When Sarah married in 1872, skirt volumes were decreasing to allow women more freedom of movement. Tiered skirts replaced hoop skirts of the early 1860s; then as skirts continued to narrow, bustles (padded with horsehair) became fashionable in the 1880s. By the 1890s, Sarah wore the more comfortable tailored shirtwaist and flared skirt, although she probably wore a corset all her adult life. Living where summer weather is frequently hot and humid made layers of clothing miserably uncomfortable. Wearing long skirts where paving was virtually non-existent created a laundering challenge.

Sarah had a large extended family with whom to exchange visits and correspondence. Ten surviving people in Sarah's sibling/half-sibling group left many trails to follow. Some of her siblings' names appear in records and family lore for a time, and then they seem to disappear. Several of her siblings left fact-filled trails. We'll look at a few of their lives for insights about Sarah's family relationships during her childhood and in her adult life.

The first child of Sarah's father's first marriage was a son born in 1829 and named Young to honor an ancestral veteran of the American Revolution. By 1850, Young lived in Baltimore where there were more varied career opportunities. By 1856, Young and his wife, Lizzie, had a daughter and three sons. With a nod to his parents, Young named his first two children Margaret and William. Although Young moved to Baltimore before Sarah was born or during her infancy, frequent trips to Baltimore during her adulthood afforded Sarah opportunities for continued relationships with Young and his family.

Hester Ann, Sarah's half-sister born in 1835, married when Sarah was five years old. Hester Ann lived nearby for a few years then moved to Baltimore as well. Six of her ten children survived her death in 1879. Although Hester's husband, John, returned Hester for burial among her ancestors in Dorchester County, he and their children continued to live in Baltimore, where John worked as a house carpenter. By 1880, their children (then in their early twenties) secured industrialized jobs such as collar maker and box maker. Perhaps when Sarah accompanied Luther to Baltimore, she visited Hester Ann's children.

One half-brother, Riley, was the youngest of the children of the first group, and he was still at home in 1860 when Sarah was ten years old. During the American Civil War, Riley fought for the Union, enlisting as a first lieutenant in the Maryland Infantry on September 1, 1861 in Cambridge, Maryland. Riley served in relative safety during the first two years of the war. His company participated in an expedition to nearby Virginia counties on the Eastern Shore, where they remained until Lee's 1863 invasion of Pennsylvania. Offering their services to support the fighting at Gettysburg, Riley's unit joined other Maryland troops in Baltimore, and then marched to Gettysburg, arriving on July 3, 1863. They joined the battle on Culp's Hill where they lost nine men during battle, and later lost fifty-two of wounds and disease.

Assuming the family knew Riley was in Gettysburg, the reports from that battle must have been horrifying. Fortunately, even after battles and seven hundred miles of marching, he survived. Riley's parents and siblings were not the only people back home worrying about his safety. On November 13, 1862, while stationed on the Eastern Shore, Riley married Emma, the daughter of a Methodist minister.

Emma's minister father had been serving as chaplain to Riley and his fellow soldiers. Clearly, the minister's family, including Riley's bride, were together in Pennsylvania. Emma and Riley must have had interludes of marital bliss during war. In September 1863, Emma gave birth to their first child, a son who only lived four days. While Emma was pregnant, her husband fought at Gettysburg, and perhaps she helped her father minister to the troops.

Typical of the era, Emma was soon pregnant again, and gave birth to her second son, Persifor, on August 27, 1864 in Pennsylvania. In addition to Persifor (called Percy), Riley and Emma added a son, Oden, to their family in 1866 (who died before 1880), and daughter, Edith May, in 1869. Riley and his family returned to the Eastern Shore where by July 1870, Riley, Emma, and their children were living with Emma's family in Snow Hill, Maryland. Did Sarah visit with Riley and his family that summer? By today's standards, sixty miles between households in bordering Maryland counties doesn't seem far. In their area in 1870, sixty miles by land was more difficult than traveling greater distances by water. Visits were infrequent if they visited at all.

Perhaps their father's October 1870 death, followed by the division of William's estate, gave Riley the financial freedom to leave Maryland. In 1871, Riley moved his wife and growing family to Indianapolis, Indiana, where Riley established a cement contracting business. Perhaps Sarah never saw Riley or his family again. Sarah's nephew, Percy, became a house contractor, married, started his own family, and became politically involved in the Democratic Party. Edith May, Sarah's niece, married a man who was active in the police department and courts of Indianapolis.

Riley and Emma added more children to their household after arriving in Indiana. Indianapolis-born children included Alice, born in 1871 (and died in 1873), and a son named Houston, born in 1875. Houston had a successful career in horticulture, ultimately managing the parks department of Indianapolis. Twins William Oden and Olen Riley were born in March 1878. Olen died in May 1878 and William in July 1878. Butler, born in 1883, later joined his father in the cement contracting business.

I imagine Sarah corresponded with Riley and knew how his family grew and prospered in Indiana. Riley outlived her by five and a half years. Sarah did not live to read newspaper stories about his later years or his obituary. After a brief illness, Riley died at the age of seventy-six in 1913. His obituary mentioned his special relationship with a local fire station. The men of Indianapolis Fire Station 14 gave a special surprise dinner in Riley's honor in March 1911. Riley had visited Station 14 almost daily for many years and founded a club called the "Firehouse Clan." The firemen reserved a special place for him to hang his coat during his frequent visits. Perhaps Sarah mirrored her brother's congenial personality traits.

Sarah had six full siblings who survived to adulthood. Enoch seems to disappear after the 1870 census. Sarah's other siblings appear in records at least long enough to determine where they were and what they were doing during Sarah's lifetime. Following their oldest half-brother, four of her brothers went to Baltimore where they worked in more industrialized jobs and raised their families. Her youngest brother, Henry, also went to Baltimore. He married Millicent Naomi who gave birth to ten children. Sarah apparently had a close relationship with Henry and his family, as did Sarah's children. As mentioned in Daisy's chapter, Henry and Millicent provided overnight Baltimore accommodations for Daisy and her children when they made their summer trips to visit Sarah.

Descriptions provided by Sarah's descendants offer clues about her home environment during her marriage. Situated on two acres of land bordered by Madison Bay and a small creek, Sarah and Luther's home featured a spacious yard heavily wooded with birch, cedar, and fruit trees, surrounded with a whitewashed picket fence. The house was small with a breezeway between the main house and the separate building housing their kitchen and dining room. Next to the breezeway was a smokehouse. Without plumbing, the family's fresh water came from a cistern situated to catch rainwater from the roof, and an outhouse was located behind the house. Between their home and Luther's store, they had a barn with a carriage house. Behind the store, Luther had a long shed for farm tools. Written descriptions were verified when Cousin Sarah, Oden's daughter, graciously provided photographs of Luther, Sarah, their home, and Luther's store. The house and store were gone before I visited Madison, so I'm grateful someone captured images of the buildings and the people before they vanished into history.

Sarah, in an undated photograph, appears to be in her forties. Her dark hair, parted in the middle and pulled back, is probably in a bun or twist out of the camera's range. Trimmed with a white lace collar, Sarah's dark dress or bodice is otherwise unadorned. Her only accessories are small earrings. She has an unsmiling, almost unpleasant facial expression.

Sarah lived long enough to raise Oden to his teenaged years. She died on July 27, 1907, at the age of fifty-seven, when Oden was seventeen. Sadly, she wasn't alive when he married, or to help him endure the loss of his sons. Oden's daughter and Sarah's namesake, was born thirty years after Sarah died.

Since Daisy often took her children to spend summers with her parents on the Eastern Shore, Sarah spent time with Daisy's children, who have spoken at length of their many visits in their grandparents' home. Sarah and Luther's next-door neighbors were a couple whose grandchildren also came for summer visits. A boy named Emerson was among their number, as you learned in the chapter about Sarah's granddaughter, Marie.

Sarah made her grandchildren feel welcome and loved. Even if Sarah found the visits tiring, she hid her feelings and warmly welcomed to her grandchildren. For her daughter and grandchildren, visits to Sarah's home filled far more than a social need. Without goods from Luther's store and food from her parents' garden, Daisy's challenge of providing for her children on a minister's salary would have been more difficult.

Sewing a year's worth of clothing for Daisy's entire family was a major task performed during Madison summers. Daisy was a skilled needleworker whose primary teacher, I assume, was her mother. During hot summers in Madison, Sarah and Daisy probably sat on the porch stitching and chatting, with their laps full of sewing. Although Sarah had only had one child at a time in her own home, having a house full of grandchildren for whom to cook and sew must have been a familiar experience to Sarah, who had grown up in a house full of siblings and half-siblings.

Food was plentiful at Sarah's table. Fresh seafood was abundant as were fresh vegetables and fruit from Luther and Sarah's garden and local farms. Children enjoyed cakes and fruit pies. Thanks to my grandmother, Marie, I have the recipe for *Grandmother's 1-2-3-4 Cake.* A typically rich, southern-style cake (only a couple of eggs short of being a half-pound cake), the ingredients include 1 cup of butter, 2 cups of sugar, 3 cups of flour, and 4 eggs. After completing instructions for the cake, Marie added a recipe for sugar cookies. Now I know the origin of the recipe for wonderful sugar cookies made by Sarah's daughter, Daisy.

Let's leave the subject of physical food to identify the source of Sarah's spiritual food. Sarah's ancestors worshiped as Episcopalians. By Sarah's time, her family worshiped as Methodists. Even after they clearly had been worshiping as Methodists for decades, the family continued to bury their loved ones with their ancestors in the cemetery of Old Trinity Episcopal Church at Church Creek near Madison. Church records of marriages and baptisms date the movement of Sarah's ancestors from the Episcopal church to Methodism. The name of Sarah's mother, Hester Ann, does not appear in Episcopal baptismal records, although the 1803 marriage of Hester's parents does appear in an Episcopal marriage record. Those records indicate Sarah's maternal grandparents began to worship as Methodists after their 1803 marriage but before Hester Ann's birth in 1818. The early growth of Methodism apparently influenced Sarah's grandparents.

Charles and John Wesley had made a mission trip to Georgia as early as 1736; however, Maryland became the formal birthplace of Methodism in America. Founded in Baltimore in 1784, the denomination sent out preachers and grew rapidly. From Baltimore, it was a relatively quick trip by water to plant a few preachers where Sarah's family lived. By 1807, a man named Joshua Thomas had been converted to Methodism, and began using his canoe (*The Methodist*) to visit Maryland's Eastern Shore and nearby islands to proclaim the gospel. By 1844, there were 4,000 circuit-riding Methodist preachers spreading the gospel.

Sarah, Luther, and Daisy worshiped at the Methodist church across from her home. Did Sarah provide the training and encouragement for Daisy's spiritual upbringing? Did their church's spiritually gifted ministers and lay teachers edify young people in the faith? Surely, a thirty-three-year-old minister needed a spiritually mature wife. Perhaps when Vincent and Daisy met and courted, Daisy's spiritual age exceeded her youthful age of sixteen. Vincent must have thought her appropriately spiritual to be his helpmate. Let's look back and see what might have facilitated Sarah's own spiritual growth to prepare her for teaching her children.

Sarah's older brother, Riley, married the daughter of a Methodist minister who had served as his army chaplain. Riley's father-in-law-to-be apparently thought Riley a spiritual man suitable to meet and marry his daughter. Sarah and Riley received spiritual training in their parents' home and their local church. There is also evidence of continued Methodist affiliation among Sarah's siblings who went to Baltimore to work and raise their families. Their marriages and deaths appear in Methodist records and publications. Their graves lie near a well-known historic Methodist church. If outward evidence indicates inward change, Sarah and her kin were believers.

Let's return to our two Dorchester County events of 1849. Harriet Tubman's life is historically significant. Her bravery was exemplary and merits praise. I also applaud Sarah who performed her tasks well by raising one son to be a good husband and father, and one daughter to become the helpmate for a godly preacher.

Sarah facilitated her daughter's marriage to a man of God. Sarah opened her home to circuit-riding preachers before and after she welcomed Vincent. If she and Luther had not invited Vincent to stay in their home and eat at their table, Daisy might have married a materially rich man yet lived in spiritual poverty. Women who follow the scriptural admonition to be hospitable might indeed be enhancing future relationships. Sharing a meal or having a houseguest might seem simple, yet scripture tells us we might be entertaining angels. God can use people we invite into our homes, and introduce to our children or grandchildren, to create alliances for His glory.

Harriet Tubman reportedly helped more than three hundred people escape slavery to enjoy freedom, and the descendants of those freed people benefit even today. Sarah trained a daughter who became a helpmate to a man who introduced people to the One who frees all captives, the Lord Jesus Christ. Those who heard the gospel under Vincent's preaching could receive salvation, escape spiritual bondage, and encourage their children and grandchildren to follow their example. Sarah's effort to raise one daughter well made an impact on history, however quietly, after all.

Imagine heaven. Sarah and Harriet are sitting side-by-side on a heavenly porch. Sarah is listening intently to Harriet's account of her painful years as a slave, the daring trips she made taking slaves to freedom on the Underground Railroad, of her fight for women's rights, and of her efforts to establish a home for the aged. If there are tears in heaven, tears of regret might slip from Sarah's eyes. If Sarah had been born a generation earlier she might have been able to help Harriet.

Some of us avoid a relationship with God because we fear He'll require more than we're willing to give. What keeps us from becoming a daring Harriet? Harriet's ninety-three-year lifespan indicates hard work and danger won't kill us. Is there oppression or cruelty you feel called to stop? If it is God's mission, He'll provide the strength and remove the fear. Imagine Sarah asking Harriet if she'd do it all again. Without hesitation, Harriet would shout, "Yes!" Then courageous, daring Harriet would tell Sarah her life was equally important to God. She would tell us every woman should be willing to accomplish her God-ordained, Spirit-empowered, role in history to the best of her ability. God can weave our individual accomplishments into the perfect tapestry of His plan for this life and beyond.

After reading this chapter, perhaps you'll want to meet Harriet and Sarah. How will we recognize them? As sisters in the Lord, Sarah and Harriet have exactly the same color of skin: a heavenly shade beyond earthly description. If we want to sit with these gals on that heavenly porch, we'll have no trouble finding them. We're women so we're willing to ask for directions. Ask Jesus, He'll show you the way.

Annie

1838 – 1912

Great, great-grandmother, Annie, had a cute nickname, inspired a son to publish glowing tributes in her honor, and owned cats. She chose meaningful names for her sons, in whom she apparently instilled faith and positive character traits. Her four sons, successful men in their personal and professional lives, produced diligent, spiritually gifted descendants.

My connection to Annie is through her son, John. Having read the chapter about John's wife, Mamie, you know a great deal about, John, a Methodist minister. John and Mamie's younger son, George, married Marie (whose chapter you've also read). George and Marie's son, John, was my father.

Because my father was absence for much of my life and was unwilling to speak of his ancestors, I knew little about his family prior to my research. I never met my grandfather, George, who disappeared in 1921. I'd love to have known Annie's son, John, who died eleven years before I was born.

Cousin Fred, who posted information about graves of our common ancestors on a website, expedited my search for Annie. A descendant of Mary Eliza, a sister of Annie's husband, Fred introduced me to another of Annie's great-grandchildren, Cousin Harry, via e-mail. Cousin Harry provided names, family history, and an introduction to his sister, Cousin Ellie. Ellie immediately invited me to her Mississippi home where Harry joined us from Augusta, Georgia. Together, Harry and Ellie made Annie come alive for me.

I commend Harry's diligent efforts to research our family in the early 1960s. Prior to the availability of on-line resources, researching ancestors necessitated travel and letter writing. During Harry's initial research, he wrote to my grandfather's brother, Robert, who responded by letter, enclosing a 1900 family photo. The first envelope I received from Harry contained a copy of that photo with each person carefully identified. For the first time, I saw the likenesses of my grandfather (George) at age eight, my great-grandparents (John and Mamie), and Annie, with members of their extended family. The clan had gathered in New York City where Annie was living with her youngest son, William, whose job had taken him north.

Our family photograph was in pristine condition and quite clear. When I first viewed it, the images appeared blurred because my eyes brimmed with happy tears. Years later, I still pause to enjoy Annie's face when I leaf past that photograph. Before we zoom in for a close look at Annie's face, let's zoom out to see her as the centerpiece of the photograph. Annie, a widow of four years, perches in the center of the family group consisting of her four sons, the wives of the three sons who were married at the time, and eight grandchildren.

Annie, sixty-two, has white hair and a slightly plump figure. Her left hand rests at the waist of her dark dress. Her right hand is open, palm up, on the lap of her little granddaughter, Anne, who appears to perch between Annie and the child's mother, Rena. Granddaughter Anne's little hands rest on Annie's palm, with Annie's fingers curled around them. Perhaps Annie was helping keep the little girl still for the photograph. I prefer to Annie held her grandchildren's hands often.

Annie has a winsome little smile on her lips and a crinkly smile in her light eyes (behind tiny spectacles). Her expression makes me wonder what she was thinking at that moment. Perhaps she was delighted her sons and their families had journeyed from Virginia and Georgia for their New York reunion.

When Annie was thirty-seven, she was nicknamed Mama Willie after the birth of her last child, William. You'll read about William as he appears in the birth order of Annie's sons. Called Willie, he became an identifying factor for his mother. After his birth, Annie was called Willie's Mama, then later Mama Willie. Even her grandchildren referred to her as Mama Willie, and she signed correspondence to them using her nickname. My cousins still refer to her as Mama Willie; to me, she is Annie.

Cousin Ellie provided a unique insight about Annie when she supplied a special photograph. Captioned "Mama Willie's Cats," the photo is a little black-and-white snapshot depicting five cats eating from a pan outdoors. We love having proof Annie's love for God's creation extended beyond people.

Cousin Bill of Maryland has done a magnificent job of capturing Annie's paternal family history. Her father's ancestor, Thomas, came to America from England in 1664 to defeat the Dutch, and help turn what was then New Amsterdam into New York. Thomas, pleased by what he found in America, stayed. After serving as a soldier and musician in the English military, Thomas married an English girl, acquired farmland, and started building a family of productive citizens. Successive men helped build schools and churches. Although arriving from England, Annie's paternal line reportedly has roots in the Netherlands. We'll follow successive generations of her paternal line from New York to Annie's own Maryland birth with hints of church affiliation along the way.

The movement away from New York began with the next generation after Thomas, the immigrant. Jacob was born in New York, and then moved to Kingston, New Jersey, where he and his Dutch wife, Jannetje, raised their family and died. Jacob served in the American Revolution, and he and Jannetje were active members of the Dutch Reformed Church.

The next generation, Jacob Jr. and his wife Ann, appear to have worshiped as Presbyterians. Their remains rest in a Presbyterian cemetery. Archives of the American Revolution capture the effect of that war on Ann's home. Ann swore to a statement of war damages, including the loss of beds, bedding, and other household articles. She conquered domestic challenges of the American Revolution while men fought for our freedom on the battlefield. Annie could be proud to carry on Ann's name.

Annie's paternal grandfather (Robert Sr.) was born in New York, raised in New Jersey, and then moved to Baltimore, Maryland. An early Baltimore city directory indicated Robert made his living as a painter. Robert married Hannah in Baltimore's First Presbyterian Church on Christmas Eve of 1797. Robert and Hannah raised their family in Baltimore, and remained there the rest of their lives. Annie's grandfather died before his wife wrote her will in March 1840. Although he might have been living when Annie was born, she would have been too young to remember him. Annie could have known her grandmother, Hannah, who died when Annie was six years old.

When writing her will, Hannah bequeathed her wearing apparel and household furniture to her two married daughters, "so that no part of them will go out of my family." I thought her words intriguing. Would her sons (one of whom was Annie's father) have allowed items to leave their family? Perhaps her sons had disposed of items left to them by their father, and Hannah didn't approve. She seems to be making a specific point when she included her "so that" phrase, doesn't she? Perhaps she had inherited treasures from her own female ancestors, and wanted those family heirlooms to pass into the appreciative hands of her female descendants.

Hannah's will goes on to request that, after paying her debts, all her cash and other property be equally divided among all her children including her four sons, the youngest of whom was Annie's father. Although not every woman wrote a will, women whose husbands predeceased them, and women who inherited land or other valuables, needed to direct the disposition of their property. Wills of our female ancestors provide clues about what (and who) they deemed most important.

Annie's father was Presbyterian during his childhood, and her mother's parents were Episcopalian. By the time of Annie's arrival in 1838, her parents were probably Methodists. Their graves are located in a cemetery associated with an historic Methodist church in Baltimore. Annie began worshiping as a Methodist in her childhood then, after marrying a Methodist man, raised her sons in that denomination.

From Mamie's chapter, you know Annie's son, John, served as a Methodist minister for half a century. Annie might have been proud of her minister son, and I'm certain she would be equally pleased when her descendants give their time and money to further the good work of other Bible-based denominations as pastors, pastors' wives, and as lay people.

We've taken a quick trip through the history of Annie's paternal ancestors from the time of their arrival in America to their Baltimore years. The family history of Annie's mother, Naomi Sophia, brings a different geographic heritage into Annie's story. Her mother's father, Andrew, was an eighty-five year old widower in 1850 when he was living in Annie's Baltimore family home with his daughter, son-in-law, and grandchildren. Quite late in my research, I found a marriage announcement for Naomi Sophia's parents stating her mother, Charlotte, was living in Fells Point, an historic section of Baltimore, and that Andrew was from Zanesville, Ohio.

Andrew's home location was a surprise since I had assumed Sophia's parents were Maryland natives, or East Coast people from a nearby state who had never lived elsewhere. It appears Andrew was born in Maryland, went west, and then returned to Maryland. Andrew and Charlotte were married on September 28, 1809 at Baltimore's Trinity Protestant Episcopal Church, where I assume Charlotte worshiped with her family at the time of her marriage. Aside from Andrew's appearance in the 1850 census, my information about Naomi Sophia's parents stops at their marriage.

Continued research might reveal more about the people of Annie's maternal family tree. So far, I've hit was genealogists call *brick walls*. Annie probably knew the history of her mother's people. Annie was twelve years old in 1850 when the census indicates her mother's father lived in their home, so Annie was old enough to have memories of Andrew as a man of eighty-five. I hope Annie had the opportunity to know her maternal grandmother, Charlotte, or knew of Charlotte's character through family stories.

With Annie's ancestry as our backdrop, let's bring Annie on stage and follow her life as it unfolds.

On August 25, 1838, Annie joined her twenty-month-old brother, George, in the home of their parents, Robert and Naomi Sophia. After Annie's birth, siblings continued to arrive in the household: Melissa in 1840, Robert Jr. in 1842, Laura in 1847, Robertina (called Bertie) in 1850, and Emma in 1854.

The August 1850 census provides clues about Annie's family life when she was twelve years old. She and three of her siblings were attending school, and her father was working as a ship's master mate. A ship's master mate (a lower-ranking officer), reported to the ship's master (responsible for navigation) and performed such tasks as standing watch, assisting with navigation, and other duties assigned by the master. Perhaps her father's sailing career facilitated an introduction to a young seaman named Isaac Henry. Did the two men work on a ship together? Did Robert invite Isaac to his Baltimore home for dinner? Was Annie's father playing matchmaker? If not, did Annie's chosen spouse meet with her parents' approval? I have no reason to believe they weren't pleased with Isaac.

On Wednesday, September 10, 1856, eighteen-year-old Annie married Isaac, a twenty-six-year-old ship captain. Annie, a city girl, was marrying a man raised in more rural Somerset County on the Eastern Shore of

Maryland. Without a portrait or photograph of Annie on the day of her wedding, fashion history helps us imagine her bridal ensemble. Baltimore women of Annie's era dressed well and followed fashion trends. Charles Dickens, an author whose prose includes clothing descriptions, visited Baltimore in the 1850s. He noted and later praised the beauty and stylish attire of Maryland women.

White wedding gowns came into vogue after 1840 when Queen Victoria wore white. Although wealthy American women adopted the style by 1856, women of Annie's social class usually wore their best bright or pastel silk taffeta dress. A typical 1856-era dress had a tight-fitting bodice and a full, tiered skirt (supported by hoops or by layers of horsehair petticoats). Embellishments for the bodice and skirt included ruffled flounces, silk flowers, and ribbons. A reticule (purse), hat, and kid gloves accessorized the ensemble. Wearing layers of cloth, scratchy petticoats, and a tight corset on a warm, humid early-September Maryland day, Annie was supremely uncomfortable.

Annie's parents announced her wedding in a Baltimore paper about six weeks after her marriage. The delayed announcement puzzles me. Did they elope? Did her parents need time to adjust prior to making a formal announcement? Annie appears to have continued a congenial relationship with her parents during her married life.

A little over nine months after her marriage, Annie gave birth to her first child. Robert Isaac, named for his grandfathers, was born on June 15, 1857, in Baltimore. His place of birth could indicate Isaac and Annie lived in Baltimore during their early marriage, or Annie needed her mother's help with her first baby.

Annie's next baby was a mystery to me until the summer of 2008. When I read the 1910 census, taken when Annie was a seventy-two-year-old widow, one of her answers puzzled me. As you read on in her chapter, you'll learn she had four sons who lived beyond Annie's lifespan. Annie had given birth to only four sons, we all believed. The 1910 census form includes two related questions: "How many live births?" and "How many children now living?" Expecting both of Annie's answers to be *four*, I was surprised to see the answer Annie gave to the first question. She answered *five* live births. After asking several cousins to search their memories for an explanation of Annie's answer, and to search their family records, no cousin could find or recall anything about a child who had died.

The answer to my question was sitting a shelf in the excellent Clayton Library for Genealogy near my home in Houston. Records of obituaries published in Baltimore included a record of the death of little "John H." The obituary mentioned Annie and Isaac, and then mentioned the child's Baltimore grandparents (Annie's parents). Apparently, Annie gave birth to her second son in February 1860 and he died ten months later in December 1860. At twenty-two years of age, she experienced the first personal tragedy of her married life.

At baby John's death, Annie's mother, Naomi Sophia, knew what to say. Annie's younger brother, Robert, had passed away in 1850 at the age of eight. Twelve years old at the time of her brother's death, Annie must have grieved with her parents and siblings; but with John's death, she understood loss from a mother's perspective and, like Naomi, endured.

As 1861 ended, the American Civil War had begun, and Annie had conceived another child. On September 13, 1862, Annie gave birth to John Thomas (my great-grandfather) in Somerset County, Maryland, where many generations of his paternal line had been born. We believe his middle name, Thomas, honored his father's brother who lost his life in the war.

The war doesn't seem to have interrupted Annie's childbearing years. She gave birth to another son, George Washington, on February 21, 1865. Since she gave birth to John in Somerset County, then to George in Baltimore, perhaps Annie's home location in those instances indicates the home port of Isaac's ships. Because of the war, and increased manufacturing in Baltimore, Isaac likely repurposed his fishing fleet to carry goods or troops.

When Annie and Isaac spent time with his mother, Mary, they lived in a place called Dames Quarter, Somerset County, Maryland. Isaac's father had reportedly been lost at sea sometime before 1850, so I assume Annie never knew her father-in-law. Little is known about Annie's mother-in-law, Mary, but facts about Mary's other children have become known through my own research, and with the generous help of Cousin Fred. Because Mary's children may reflect their mother's spiritual values and character traits, we'll look at Fred's ancestor, Mary Eliza.

Annie's husband, Isaac (born in 1829) was the eldest of three boys and two girls. Isaac's youngest sister, Mary Eliza (born in 1843), was the baby of the family. Mary Eliza lived in her mother's household until her marriage on May 20, 1863. A lengthy tribute, written at the time of her death in

1891, describes Mary Eliza as a woman who lived by God's word, and dedicated herself to her husband and children. She was diligent, hospitable, and uncomplaining. Even though ill for fifteen years, partially deaf, unable to walk, and homebound for seven years prior to her death, Mary Eliza remained patient and confident in her faith. Since Mary Eliza's illness began during the last four years of her mother's life, Mary may have helped care for her daughter and grandchildren. If Mary Eliza reflected the character of her mother, Annie married into relationships with strong women of faith.

During the war, Annie dealt with shortages, as did her peers. She might have dyed her dresses when they faded and fabric was scarce. When fabric appeared worn, perhaps she took clothing apart and turned it inside out then re-stitched the pieces. The availability of food was an issue for many people during the war. Could Annie better feed her family on the Eastern Shore or in Baltimore? By 1870, Annie's father was working as a Bakery Superintendant, a profession in which he and his son, George, participated for many years. If her family had begun to work in the bakery business in the 1860s, they might have been certain they'd have bread in Baltimore if they could get flour. The Eastern Shore's seafood and crops could provide a fresh, healthy diet during wartime shortages elsewhere. Crowds of soldiers and other travelers increased the threat of disease in already densely populated Baltimore. Perhaps Annie lived alternately in both locations during the four years of war.

What were the psychological and emotional issues for Annie during the war? Maryland, where she was born and lived until after the war ended, was rife with divided sentiments. Reportedly, more than 62,000 Marylanders fought for the Union, and 22,000 fought for the Confederacy. Her father was not a farmer and owned no slaves. Her husband's family had been watermen for several generations. I assume they sided with the Union.

Annie's family had a history of anti-slavery sentiments. A Baptist minister in Annie's paternal ancestry took his views to the pulpit. Still quoted in contemporary texts on the early abolition movement, his 1773 sermon suggested slaves rebel against their masters and emphasized the eternal penalties due people who violate the rights of others.

Isaac and Annie somehow endured the war and the beginning of reconstruction in Maryland, and then moved to Norfolk, Virginia in

1869. They moved to settle in Norfolk because Isaac had begun a fertilizer business (then called *guano works*). He continued his earlier vocation, however, because census records and Norfolk Directories alternately described his profession as sea captain, then as a factory superintendent of guano works. He apparently earned enough money for housing, clothing, and food as well as to provide educational opportunities for his sons. The family worshiped at Epworth Methodist Church in Norfolk.

Annie's thirteen-year-old son, Robert Isaac, appears in the Baltimore household of Annie's parents during the census taken in June 1870. Annie's father was working as superintendent of a bakery; a business Annie's older brother, George, entered as well. Robert Isaac's introduction to the food industry that summer later influenced his career choice.

In the opening of Annie's chapter, I mentioned her Mama Willie sobriquet, created because of the late birth of her last child, William. At thirty-seven, Annie wasn't beyond childbearing age when William was born. What must have seemed remarkable was the ten year and one month delay between sons George and William. She had given birth to George in February 1865, and then she gave birth to William in March 1875. Annie must have given away all the baby furniture and little boy clothing used by her earlier sons, and had to furnish her nursery anew. To Annie's advantage, her older boys were seventeen, thirteen, and ten when William was born, so not only did Annie have plenty of experience raising sons, Willie had three older brothers to keep him entertained.

The 1880 census indicates Annie's eldest son, Robert Isaac, was still living at home, twenty-two years old, and working as a bookkeeper. John was then eighteen, George was fifteen, and William was five. To lighten Annie's 1880-era workload, she had a live-in maid. According to Norfolk directories and census data, Annie and her family moved every few years. They may have moved to better neighborhoods as their income grew, or they moved to smaller quarters as their sons began to leave home.

Isaac's mother, Mary, passed away on February 15, 1880. With Mary's death, Annie's in-laws were now both deceased, but her parents survived a few more years.

Annie's nest began to empty in the 1880s. Robert married a woman named Annie Irene "Rena", on December 21, 1882. Rena and Robert gave Annie seven grandchildren who, by all accounts, Annie adored. They asked Annie to choose names for six of their seven children. Naming and enjoying female grandchildren must have been delightful for Annie, a mother of four sons. Annie honored her mother, Naomi, when the first girl arrived in 1883. The second child, Roschelle, honored the child's maternal grandmother. On May 17, 1888, Annie chose the name Irene May for a girl they identified by her middle name and birth month, May.

When Robert Isaac and Rena's fourth child and only son arrived on July 27, 1889, naming was easy. Annie chose Isaac Henry, the name of Annie's husband (the child's paternal grandfather), and the child's call name was Harry. Rena had three more girls: Anne (called Annie) was born in 1897. When the next granddaughter was born in 1901, Annie suggested naming her Ruth. Cousin John says family lore mentions Annie thought Ruth the perfect name for a child whose older sister bore the Biblical name of Naomi. Perhaps the book of Ruth was among Annie's favorite Bible passages. Then Annie gave up her naming job saying her son should name the last child. I imagine Annie was pleased when her son gave his last daughter the call name of the wonderful woman who had born all his children: Rena.

My great-grandfather, John, received the call to preach when he was a young man attending Epworth Methodist Church in Norfolk. He went to seminary at Randolph-Macon to prepare for the ministry, began his pastoring career in 1884, and married in December 1888. The two grandchildren Annie gained from that union were Robert, born in December 1889, and George, born on Christmas Eve 1891. You read John and Mamie's life stories in Mamie's chapter of this book. From Annie's perspective, her preacher son moved often as the Methodist church reassigned him. Annie joined the churches John pastored when they were close to her home; however, during most of his career, John's church was too far away for Annie to attend his services. She heard him preach during brief visits to John and Mamie's homes in various Virginia and Maryland locations.

Although third in birth order, Annie's son, George, married two months before his older brother John. On October 23, 1888, George married beautiful Elsie in her parents' Norfolk home. Annie must have been present and smiling when her second son, John, performed the marriage of her third son. The announcement of George and Elsie's wedding called

George a well-known young citizen of the city of Norfolk. Following their marriage, George and Elsie lived in Augusta, Georgia where their daughter, Cecile, their grandson, and great-grandchildren were born and where several descendants still live. George succeeded in the cotton and banking businesses to support his family.

George's family grew spiritually in a Baptist church and taught their descendants to be people of faith. From one daughter, George and Elsie's descendants include wonderful business people (one of whom is an active Episcopalian), as well as a Presbyterian Church in America (PCA) pastor, and two PCA pastor's wives. Annie must be smiling in heaven over the positive impact of George, Elsie, and their descendants. Cousin Ellie (who inherited Annie's love of cats) descends from George and Elsie, as does her brother, Cousin Harry, who graciously shared his early genealogy project.

Annie's daughter-in-law, Elsie, was a sweet, quiet woman known for her beautiful knitting and crochet work, and her impeccable manners. Elsie's great-granddaughter, Ellie, suggests (laughingly) that Elsie was quiet because her husband talked incessantly. George and Elsie had only one child, Annie's lovely granddaughter, Cecile. Cecile's granddaughter, Ellie, still owns the book *The Age of Fable,* inscribed by Annie with the words: "To Cecile from Mama Willie, Christmas 1900." Please note: book-giving grandmothers live longest in our memories.

Let's travel back to Baltimore to catch up with Annie's siblings and parents. Annie's decade-younger sister, Robertina, married a man named Ezekiel in 1873, and had three children. Before 1880, Ezekiel and two of their children died, and in 1881, the third child died. When the 1900 census taker asked about Robertina's children, she reported having experienced three live births, yet she had no surviving children in 1900. How Annie's heart must have ached for her sister.

Annie's sister, Laura, was four years her junior. Laura married a first cousin named Charles, and then she passed away in the 1880s leaving her husband and several children. Laura's family history illustrates the weakened offspring produced by close intermarriage. Of the eight children born to Charles and Laura, only four reached adulthood. When one of the surviving daughters of Laura's household had five children, and all five died in their youth.

Adding to the sadness of losing nieces, and nephews, Annie's mother died on July 16, 1883. She was working in her Baltimore home when she suffered a stroke and suddenly passed away. Thanks to Cousin Bob who still lives in Baltimore, Naomi Sophia's obituary was located. The words penned by her family provide clues about the woman whose character helped shape Annie's personality.

Naomi Sophia was a consistent Christian, ready to help others, and devoted to her family. Her obituary, written by her husband and children, went on to include a poem, claiming Naomi Sophia would rise in victory when Christ returns. Since publication of the tribute occurred on July 21, 1883, Annie had a few days to travel to Baltimore to help her family write their thoughts about Naomi. Annie's father passed away in May of 1890.

Events were occurring among Isaac's siblings as well. Mary Eliza, Isaac's baby sister, died on December 27, 1891. In contrast to Annie's long, healthy life, her poor sister-in-law died at the age of forty-eight after a fifteen-year illness.

What sort of man was Isaac and what did he do in addition to his career? I've never seen his face. He died prior to the 1900 family photograph, and doesn't appear in other family photographs. No stories about his personality have surfaced. Interestingly, I learned of Isaac's furniture building skill when Cousin Butch (a descendant of Annie's oldest son, Robert Isaac), provided a photograph of a beautiful chest of drawers handcrafted by Isaac. We have hard evidence Isaac created at least one heirloom, yet we don't know how extensive Isaac's woodworking activity might have been.

Isaac may have modeled the great work ethic exhibited by his sons, although Annie's family members also exhibited diligence and success. Based on census records, city directories, college records, and family history, there was no idleness in Annie and Isaac's household. Since Isaac's sons did not work at sea or in the fertilizer business, Isaac apparently did not push his vocational choices on his sons. They chose and excelled at their varied professions.

Isaac's ancestors came to America early and, like Annie's ancestors, were working class people who helped build the nation. Their ancestors had much in common. William, Isaac's Dutch ancestor who came to America from England, first lived in Northampton County, Virginia in 1647. In the early 1660's, William moved the family seat up to Somerset County, Maryland. Apparently known and respected in Church of England circles, early Maryland history indicates William held church meetings in his home before church buildings were constructed. He chastised neighbors who did not baptize their children into Church of England membership or failed to pay their tithe. As directed by early Maryland government, he jailed men for holding Quaker meetings in their homes. Ironically, those Quaker men he jailed produced daughters who later married William's sons, grandsons, and great-grandsons. The history of our family mirrors the story of our country's early struggles to ensure the status so many came here to achieve: religious freedom.

William, who died in 1665, and the men who came after him, farmed their land and sailed their boats. They served in such roles as land commissioner and high sheriff. They became guardians for orphans in their midst, worked as skilled tradesmen who trained apprentice craftsmen, organized militia groups to defend their land, and one man wrote an early history of Maryland.

Isaac died of liver cancer on May 15, 1896. He must have been hospitalized or living in nursing home, because the 1894 Directory for Norfolk lists Annie living alone in a boarding house. For several years after Isaac's death, she appears to have lived in her own home on Bute Street in Norfolk.

By 1910, Annie's youngest son, William, was working for Robert Isaac in his Norfolk peanut products business. On June 26, 1912 (a few weeks before Annie passed away), William married Minnie, secretary-treasurer of Robert Isaac's company. Annie must have been happy she lived long enough to see her youngest son marry. Minnie had been closely involved in Robert Isaac's business for a decade, so Annie had ample opportunity to know Minnie well, even though she had only weeks to love her as a daughter-in-law. Minnie and William had twenty-one years together, and then Minnie died suddenly in 1933. William died at the age of seventy in the home he was sharing with his sister-in-law, Rena.

Since William left no descendants and our family knew little about Minnie, I searched for descendants of Minnie's siblings. I'm so grateful to Minnie's descending nieces and nephews (by blood and by marriage) who graciously spent time sharing information about Minnie's personality. Minnie was an artistic gal who left lovely porcelain pieces and canvas works she painted as a hobby. One of her great-nieces remembered happy visits to William and Minnie's home.

Ancestral and historic naming conventions are readily apparent in the names of Annie and Isaac's older three sons. The name chosen for their last son, *William McGee,* was puzzling. Was William McGee a person Isaac and Annie knew? Was William McGee a man they admired from afar? I found the answer. William McGee was a Methodist minister who pastored churches in Norfolk while Annie and her family lived there.

William McGee spent his childhood in a Christian denomination that did not emphasize Bible study. As a young man and confirmed unbeliever, he happened upon a Methodist service where he heard the Biblical story of Daniel. After hearing that sermon, William wanted Daniel's God in his life. William accepted Jesus, began attending Sunday school, and then began to teach. He became a beloved preacher, led churches in several Virginia locations, and was in the Norfolk area for five years prior to the birth of Annie's last son.

Perhaps Annie admired William McGee's preaching or loved him for his Christian character. Was the name a good choice for Annie's son? William did worship as a Methodist and apparently lived an upright life. My research revealed an even more intriguing byproduct of the family's relationship with Rev. McGee. By naming her son after Rev. McGee, Annie may have created an alliance that would benefit her older son, John.

In 1885, ten years after Annie created his namesake, Rev. McGee was pastoring Asbury Methodist Chapel in the Atlantic City section of Norfolk. The following year, Rev. McGee was pastoring another Norfolk church, and the pastor at Asbury Chapel was Rev. John, Annie's older son who had entered the ministry in 1884. Since Atlantic City was John's first charge, he must have been training under Rev. McGee in 1884 and 1885. Annie's friendship and honor toward Rev. McGee established a relationship before 1875 that provided an excellent mentor for her minister son, John, a decade later. God and mothers work in mysterious (wonderful) ways.

I've mentioned the highly successful food business of Annie's son, Robert Isaac. A few years after the initial introduction of peanut butter, Robert Isaac and his partner, who had been in a successful salted nut and peanut candy business, began producing peanut butter commercially. According to his descendants, Robert and his partner were the first to commercially market peanut butter in the United States. After years of highly profitable success, they liquidated the company in 1924 and Robert died in 1926.

Robert Isaac merits praise. Although many of his accomplishments occurred after Annie's death, she would have been proud of her son, his wife, and their descendants. Census records and city directories document Robert's diligent approach to every task and career move from youth to retirement. Best remembered as a partner in a lucrative peanut product business, I was further amazed to learn he also served as treasurer in a pile driving, warehouse building, bridge and wharf construction business formed by his wife's stepfather.

Robert's wife and large family enjoyed a beautiful home in a lovely neighborhood, where his wife had help with keeping house and raising their seven children. Descendants say Robert and Rena employed a household staff consisting of a cook, maids, and a chauffeur. Apparently never selfish with their hard-earned prosperity, Robert and Rena's hands were always open and giving. They invited sailors stationed at the nearby naval base to their home for family-style meals and warm friendship. Perhaps Annie encouraged her son's hospitable nature. His greatest accomplishment was choosing an excellent wife.

Before and after Robert's death, Rena appears to have welcomed members of Robert's family into her home. In Mamie's chapter, you read about my great-grandmother and others living with Rena after their spouses died. Rena didn't permit loved ones to be hungry, lonely, or homeless. It is no surprise to me that Robert and Rena's descendants mirror their giving spirits.

Annie's son George was quite a character who had lived far beyond the lifespan of his parents and siblings when he died at age eighty-nine. George proved working keeps people young. When he celebrated his seventieth birthday, he told people he retired at fifty-five, didn't enjoy retirement, and returned to work. When he celebrated his seventieth birthday in February 1935 at the bank where he worked, he received flowers, cakes, telegrams and kind wishes from dozens of coworkers and customers. He said having a kind disposition and friendly attitude toward others produced happiness.

George developed habits for maintaining youth. He rose at five in the morning, jumped into a tub of cold water, and whistled or hummed a tune while dressing. He recommended getting to work early, eating carefully, and getting lots of sleep. He thought work the best tonic, and advised people to keep busy. Perhaps George mirrored Annie's health and work ethic training.

Four children would constitute a small family in Annie's era. Even if limiting their family size wasn't intentional, Annie's four children were beneficiaries of more of their parents' time, and more money for such expenses as college. Wise women have the number of children they are able to support in all the important aspects of life.

Annie died of uremia on the day before her seventy-fourth birthday. The last sight Annie saw were the faces of her sons Robert, John, and William who had gathered at her bedside. George, en route from Augusta, must have been disappointed to miss the last moments of Annie's life. In his expressive poems of love for his mother, George referred to Annie as *God's own best.*

Annie's funeral took place on Sunday, August 25, 1912, at her church, Park Place Methodist Church in Norfolk. Buried next to her husband, Annie's position echoes her central placement in the 1900 photograph. The remains of many descendants surround Annie's grave. Her obituary described her as a woman of many beautiful and attractive traits of character who was widely known and universally esteemed.

How did people who knew Annie describe her? Her son, George (whose sobriquet in later life was *The Apostle of Sunshine)* frequently penned his opinions. He characterized Annie as, "best friend, good, sweet, one-of-a-kind, peaceful, and serene," in a lengthy Mother's Day poem later published in an Atlanta, Georgia newspaper. George included the phrase, "eyes in whose depth lingers a smile," when he described Annie. Those words made me think of the smile in Annie's eyes I had observed in the 1900 photograph. He wrote of her sacrifices, her deep love, and her heart of gold. George praised Annie for trusting and believing in her children without restraint.

Annie made sacrifices without complaint, and could always see something worthwhile in her children, according to her son, George. Annie's positive reinforcement helped her sons succeed in all aspects of their lives. George went on to write that Annie was always ready to comfort others. He made several statements in his writings indicating he believed Annie's love to be a gift from God because her love had divine attributes. I wonder if he knew *Anne* actually means *grace.* Is there any greater gift than God's grace?

Annie raised her sons as Christians in a Methodist setting, and evidently provided early and ongoing spiritual and character training at home. Annie's job was to give her children a spiritual foundation. Evidently, she accomplished her mission; her descendants worship in Bible-based denominations.

The character traits mentioned in the obituary of her preacher son, John, more closely resemble those acquired from a mother's lessons than during seminary instruction. "He will be longest remembered," John's obituary read, "for the kindness, sympathy, and sensitive human understanding with which he discharged his duties as a minister, citizen, and friend, and which made his long life in Norfolk a healing ministry both within his own church and without." Through John's ministry, perhaps Annie's faith and love touched others.

When you think of Annie's boys, remember peanut butter, cotton, banking, hospitality, and Godly preaching. Her sons spent their careers producing products, performing services, or speaking words to meet people's needs. Annie's sons seemed more concerned about the diligence and charity they exhibited in their careers and lifestyles than in the specific professions they pursued. Mothers are strategically positioned to instill positive character qualities in their children. Annie's sons appear to have been men of good character.

Several years ago, Cousin Ellie, wife of a Presbyterian Church in America (PCA) pastor, whose children are also involved in ministry, taught a marvelous *Kinsman Redeemer* lesson based on the Book of Ruth. Listening to a recording of Ellie's teaching, I wanted to pass headphones to Annie in heaven so she could listen to her Godly great, great-granddaughter. If Annie prayed for future generations, she must be thrilled at the spiritual fruit Ellie's branch continues to produce.

Think back to Annie as the centerpiece of our 1900 family photograph. Annie definitely has a twinkle in her eye in heaven if those posed around her in 1900 now surround her for eternity. Look at a family snapshot. Will all your loved ones be assembling again in heaven? Are you certain? If they are still living, pray for them. Find non-confrontational ways to ask about their relationship with the Lord. Don't let any dear ones slip away before you're as certain as possible they're heaven bound.

What are my personal feelings about Annie after studying her life? Beyond admiring her strength to endure life's trials with grace, I think of Annie as positive and approachable. If we want our lives to be a positive influence to others, the first step is to be someone with whom people want to spend time. Annie died thirty-five years before I was born so my only hope to know her is to find her in eternity. To me, the laughter in Annie's eyes reveals a glimpse of the Lord's joy coming from her core. Annie's is not a smile based on life being good. Rather, Annie's life and countenance encourage me to realize the value of women who remain positive whether life is good or not. Isn't that the precise definition of a spiritually strong woman?

Annie's son, George, ended his Mother's Day tribute poem by describing Annie's love as a, "priceless possession that is divine and will last beyond the calendar of time." No matter how great a gal Annie was, I don't believe she left so lasting an impression on her children simply by operating under her own steam. We don't need Bible annotations or personal journals documenting that she sought God's help; we see evidence of her spiritual strength in those she trained and loved.

May we all ask God to give us the strength and desire to exhibit His joy. We need to prevent negative ideas from entering our minds by studying God's enlightening instructions in the Bible. Even when we read Bible passages we've read many times before, the Holy Spirit gives those passages fresh meaning, perfectly tailored to our need. We're encouraged in Philippians 4:6 to pray about everything. Did Annie pray for her own guidance, for her sons, and for future generations? I believe she did.

Let's take our cue from Annie (and your own smiling ancestral women), and work to help the next generation become God's best. Wars begin and end. Financial downturns eventually turn up. The value of instilling faith and ethics in children and others with whom we associate occurs on a higher plane than the everyday issues of life. With God's help, women create leaders who prevail and people who endure. Think of Annie and be a positive, Godly woman whose influence blesses many generations in this life and helps ensure the growth of heaven's population in the next.

Therese

1832 – 1908

Marie Therese, born in June 1832 in Norfolk, Virginia, was a first-generation American. Named to honor her father's sister, and called Therese, she was the first daughter born to her parents, Yves the Frenchman and Mary the Scot. Before we begin Therese's story, we will meet her parents and explore their ancestral roots to inform our view of the diverse geographical, cultural, and spiritual backgrounds of her ancestry.

Therese's father, Yves, arrived in Norfolk about 1819 and became a naturalized American citizen in 1827. Born in L'Orient, France in 1798, he attended a Catholic college in St. Malo, a port of departure for French explorers. By 1816 when Yves was attending college, he knew of French involvement in the American Revolution. French soldiers and sailors who had come to our aid returned to France telling stories of our land of freedom. America must have sounded good to Yves because he came and he stayed.

Therese's mother, Mary, was born on July 3, 1803, in Greenock, Scotland. The oldest of seven children, Mary and her family came to America shortly after her brother, Archibald, was born in Greenock in August 1814. The family had settled in Norfolk, Virginia, by 1816 when Mary's parents, Nathaniel and Elizabeth, joined the First Presbyterian Church of Norfolk. On May 27, 1817, Nathaniel became a naturalized citizen.

Immersed in Scottish speech patterns for the first decade of her life, it is likely Mary could understand her Virginia neighbors better than they could understand her heavily accented English. Assuming Yves mastered

English, he undoubtedly spoke his second language with a French accent. Imagine listening to conversations in their Franco-Scottish household. Therese and her siblings must have spoken with a blending of Scottish and French speech patterns.

Did Yves teach his children French? The children must have become fluent French speakers and writers as a way of retaining the family's heritage and for practical reasons. Members of Yves's family visited America, and Yves wanted his children to be able to communicate with their relatives. Yves Jr. returned to France to be educated where his father had attended college. Yves must have prepared his son for a French higher education by teaching him the language.

If Catholic Yves and Presbyterian Mary were at odds when deciding a family denomination, Mary's Presbyterian roots appear to win at the outset. According to a brief newspaper announcement, a Presbyterian minister wed Yves and Mary on the evening of Saturday, December 30, 1820. The name of Mary's father, Nathaniel, appears in the Norfolk marriage record as bondsman.

Born about 1825, Yves Jr. is the eldest child I've identified in Yves and Mary's family. Based on an 1820 wedding, I expected Mary to bear children prior to 1825, as well as between the birth of Yves Jr. and the birth of Marie Therese in 1832. Although the 1830 census indicates there was only one five-year-old male (Yves Jr.) in the household, perhaps there were babies who didn't survive. After Therese was born, six more siblings later joined the family: Mary Eliza in 1834, Virginia in 1837, Robert in 1839, Thomas in 1841, Charles in 1842, and Fanny in 1848.

To support his growing family, Yves worked as a fabric dyer and tailor. Fabric dying was a thriving business in the mid-1800s. Women who could not obtain fabric, or could not afford to purchase the yardage required to create dresses of the era, had their previous season's dresses dyed so they appeared new. Yves provided well for his family, so Therese enjoyed a financially comfortable childhood.

Let's imagine life in Norfolk when Therese was born in June 1832. Norfolk was in the midst of a cholera epidemic during the month of her birth. Thankfully, the family appears unscathed by cholera, yellow fever, and smallpox epidemics that claimed many lives in their city. During the first decade of Therese's life, the paving of streets, better water quality, and the construction of schools, churches, and a public library improved life in Norfolk.

By 1840 when Therese was eight years old, visitors to Norfolk described the town as a place of scenic beauty. Letters of the era mention gardens filled with azaleas, and native flowering trees such as dogwood and redbud. If Therese enjoyed a garden as part of her own childhood home, perhaps it was a small one. It appears the family lived in the heart of the city where Yves could be near his business and the family was close to shops, schools, and church. Speaking of church, as I've already hinted, the religious backgrounds coming together in Therese's family were as different as the native languages of the French father and Scottish mother.

Catholics and Presbyterians had established congregations and erected buildings in Norfolk before Yves and Mary began married life. The Presbyterians had taken an early lead in Christian education in 1816 by opening the first Sunday school in Norfolk. According to family lore, Mary put her Scot foot down and insisted her children be raised Presbyterian. Norfolk churches considered music important; their instruments and musicians were reportedly excellent. Therese loved music and perhaps her first exposure was in church. From the 1820s, Mozart was popular at indoor and outdoor musical performances in Norfolk.

The secular education Therese received began in her home. She continued her education in public or private schools. Therese became an avid reader. We have evidence she read the Bible, her prayer book, women's magazines, and novels. Perhaps influenced by her mother's county of origin, Therese enjoyed reading Sir Walter Scott's Waverly series.

We knew Therese wrote beautifully by adulthood, because her descendants inherited prayers and journal entries she penned. Thanks to Cousin Linda, I learned Therese enjoyed writing earlier in her life. Apparently, she created family gifts by employing her writing talent. What Linda graciously provided was the text of a poem Therese wrote for her father on his fifty-seventh birthday, when she was twelve. As you'll read in the poem, Therese was almost thirteen, and expected her father to return the gift in kind the following month. Before you read the text, you'll need to be enlightened about her use of an odd word. I had to head for the dictionary to learn the meaning of *ween*. Considered archaic because it is no longer in use, *ween* meant to suppose, guess, or think. Her poem follows in its entirety.

To Papa

Happy happy happy day
This is the 24th of May
The anniversary of your birth
When first you saw the smiling earth.

It was last year about this time
That I essayed to make a rhyme
In honor of that happy day
When daughter, son and all are gay.

Your age is fifty-seven I ween
And soon I shall be just 13.
From you in June I shall expect
Some poetry but more select.

Therese
May 24th 1845

In 1846, Yves Jr. returned from college in St. Malo, France. In March 1847, he joined his fellow American countrymen departing by ship from Fort McHenry bound for war in Mexico. Sadly, he did not return. Yves Jr. died when a cannon ball penetrated his body. Many men who fought in Mexico died of yellow fever. If he had to lose his life, dying quickly on the battlefield was preferable. It is doubtful they returned his remains to Norfolk; however, his name is included on the family obelisk at Norfolk's Cedar Grove Cemetery with the words, "Born March 16, 1825, died Vera Cruz, August 10, 1847."

How did Therese's family view the loss of her oldest brother? Yves, whose parents endured the French Revolution, and Mary, familiar with Scottish rebellion against the English, would believe a man's duty was to support his homeland. The loss of their eldest son, however noble they believed the cause, must have been difficult for Therese and her family to bear.

Therese, then fifteen, became the eldest child in her household. Her mother was busy raising a growing family. Therese formed close bonds with her younger siblings, illustrated when she opened her home to them and to their families in later life.

On October 16, 1849, seventeen-year-old Therese married Anthony, called Tony, a son of French parents. Yves's business was lucrative according to the 1850 census record. He estimated his real estate holdings to be valued at $30,000. Therese, daughter of a successful tailor, must have worn a lovely dress for her wedding. Perhaps Yves also provided an elegant trousseau, including all the clothing his daughter needed for her first year of marriage. For Yves, seeing Therese marry the Catholic son of fellow French immigrants must have softened the pain of seeing his young daughter leave home. He was gaining an American-born son-in-law who, it is likely, also spoke French.

How did Therese, a Norfolk girl, meet and marry a Richmond man? Absent a romantic story of Therese and Tony's first meeting, I studied their families and the events of their time. To understand the social atmosphere and personalities of Tony's family, we need to look at the lives of Tony's parents, Antoine and Catherine. Both of Tony's parents descended from French society families who experienced radical change during the French Revolution.

Antoine was a boy of seven when his wealthy 35-year-old father, Bartholomew, a member of the king's guard, was taken out of their Lyon, France home at midnight, tied to a tree, and shot. Antoine had watched as their family home burned to the ground, destroying much of their property. Antoine's mother, Marie, was the only daughter among twenty-one children of the governor general of a French Province. Of her twenty brothers, several served as officers in the army and navy, several held positions in government, and a few became Roman Catholic priests (one of whom rose to the level of bishop).

After his father's death, Antoine grew up with his mother and a stepfather, whom he disliked. To get away from the man Antoine referred to as his rich but miserly stepfather, he went to Paris to live with an aunt, whose elegant home he thought paradise. In Paris, Antoine met Catherine, his future wife. Catherine was born on March 17, 1795 on her father's ancestral estate in Metz-Lorraine, France. Catherine's family members were involved in court life; most notably, her brother served as private secretary to the king and her nephew served as secretary to the queen consort. Her family religious background was Roman Catholic as well.

Catherine (sixteen) and Antoine (twenty-five) married in 1811 and began their married life in Paris. Antoine and his sister's husband co-owned a large lamp factory in Paris, and received a contract to perform the metalwork to create the oil lamps used to light the city of Paris before gas lamps became available. Antoine was working hard and making a fortune when he decided to go to New York with prospects of returning to France in a few years with even more wealth. He hoped to win the contract to light the city of New York.

I haven't found Antoine and Catherine on a ship's passenger list; however, I calculated their arrival in New York by noting the birth locations of their children recorded in later census records. Antoine and Catherine left France after the birth of a daughter in 1815, but were in New York for the birth of a son in 1817. Antoine opened his business at 65 Chatham Street in New York City, and they lived near New York's City Hall. They remained in New York until after the birth of another son in 1820. Unfortunately, Antoine joined with a new partner who reportedly caused their firm to lose a great deal of money. Efforts to win the contract to light New York had failed.

Timing, as usual, is everything. Antoine, frustrated with his situation in New York, received a contract that changed the course of his life and the lives of his descendants. The contract to install the copper roof covering on Richmond, Virginia's City Hall gave Antoine his first glimpse of a southern city. Richmond must have been more appealing to Antoine than their location in New York. He soon moved his family to Richmond where a daughter was born in 1822, followed by Tony in 1824. Six more children were born in Richmond where the family seat remained for Tony and his siblings as well as for many individuals of future generations.

Tony grew up in Richmond where his parents were finding their way into Virginia society. As relative latecomers, entering the society of native Virginians might not have been easy, but Tony's father apparently knew Thomas Jefferson. Perhaps they met when Jefferson was in France. It wasn't long before Antoine had contracts for work on the University of Virginia at Charlottesville, and on Virginia plantations of people in Jefferson's social circle.

The cornerstone for Central College (later the University of Virginia at Charlottesville) was laid on October 6, 1817, about the time Antoine arrived in New York. Antoine was still in New York when Jefferson chartered the University of Virginia on January 25, 1819. By 1824, the former President was preparing instructions for selecting appropriate faculty members, and Antoine was in Virginia. In March 1824, Catherine was pregnant with Tony when Antoine traveled to the University site to negotiate a contract for covering the roof of the rotunda with copper. Explaining the selection of Antoine's company, a note from the project manager to Mr. Jefferson complimented the high quality of Antoine's work.

Correspondence also indicates difficulty in conversing with Antoine, apparently still struggling to master English. A note dispatched to Mr. Jefferson mentioned Antoine's upcoming arrival from Richmond, and requested Mr. Jefferson's attendance at a meeting in Charlottesville. Apparently, the presence of Mr. Jefferson at a prior meeting (because he could converse with Antoine in French) had facilitated a successful conversation.

Mr. Jefferson's response indicates he probably did meet with Antoine, but at Monticello. Mr. Jefferson, then eighty-one, penned a note explaining his last round trip to the University (without getting off his horse, he wrote) had left him so tired he could barely reach home, and so sore he could not ride again for several days. Mr. Jefferson went on to explain that if consultation were necessary, the gentleman managing the construction and Antoine should take a ride to Monticello at their convenience.

Social events surrounding the Marquis de Lafayette's visit to Virginia in 1824 provided a possible venue for Tony and Therese's parents to meet. A few weeks after the university dome covering was completed, the much-acclaimed fellow Frenchman visited and admired the results. In November 1824, Lafayette was making his way to visit Mr. Jefferson at Monticello, accompanied by a long procession. Lafayette arrived at Monticello on November 4, and enjoyed Mr. Jefferson's hospitality until November 15. Lafayette's record of the visit indicates that he learned Mr. Jefferson was building a university, and said that on November 8, they visited that institution. Perhaps someone told Lafayette the roof had been the responsibility of a fellow Frenchman. Lafayette and his entourage attended a public banquet in Charlottesville, where Jefferson toasted Lafayette for his contributions to our country during the American Revolution.

Were Antoine and Catherine members of the entourage following Lafayette's travels through Virginia? I believe Antoine was too busy running the business, and Catherine was too busy having children. Perhaps Antoine called upon his French compatriot at Monticello, or attended the public banquet at Charlottesville. Catherine gave birth to her fifth child, Tony, in August 1824, so I'm not certain she was ready to endure a rough carriage ride to Charlottesville. Were Frenchmen, now adopted Virginians, invited to greet Lafayette? If so, did Therese's parents attend as well? Did the parents meet that early to establish a friendship that later extended to their children? This scenario seems possible yet less plausible to me than the next one I'll present.

A later scenario that could have brought Therese and Tony together involves the possibility of Antoine's business providing products or services to Yves's dyeing establishment. Known for their work to cover roofs with copper, for their expertise in the use of zinc and iron to create gutters, and other metalwork including decorative ironwork, Antoine's company was also developing pipe fitting and plumbing skills.

Yves needed large metal vats for the dying process as well as plumbing to bring water to the vats, and then drain away the dye bath afterward. Wouldn't a Frenchman in Norfolk hire a fellow Frenchman of Richmond to supply the required products and services? Perhaps Antoine visited Norfolk for a business meeting with Yves, and dined with Yves's family in their home. Over the dinner table or in the parlor, Antoine could have met a girl who was attractive, bright, and could converse in French. On the way home to Richmond, perhaps he thought Therese a good match for his son, Tony. Did the crew traveling from Richmond to Norfolk to deliver and install metal products include a young man named Tony, who had already heard about the daughter of his father's fellow Frenchman? Although it is only speculation, this second idea seems plausible.

Before he married Therese, Tony was working in his father's business, where he spent his entire career. They were gas fitters during Richmond's

1846-1851 oil-lamps-to-gas-lighting conversion, so perhaps they took part in that effort. The business continued to expand to become one of the largest metalworking businesses in the South by the time of the American Civil War. Later, their company became the largest southern supplier of plumbing hardware and services during the era when homeowners were eager to install indoor bathrooms.

As early as 1849 when Tony proposed marriage, I'm sure Therese and her parents believed his family's business meant Therese had made a good match financially. They were both children of Frenchmen who had come to America and worked hard to establish growing businesses. The two young people had much in common.

As always, every story has positive and negative (or at least challenging) aspects. As the story of Therese as a married woman begins, two issues could have given the bride's parents pause. First, staunchly Presbyterian Mary was possibly distressed that Tony's family had adhered to their Catholic roots. Therese did join Tony in the Roman Catholic Church at their marriage. Printed in the Richmond Inquirer, their wedding announcement stated that the Rev. Mr. Histelberger married Tony and Therese on Tuesday, October 16, 1849, in Norfolk. Research revealed that Rev. Alexander Histelberger was a priest of the Norfolk Catholic congregation at the time. There was no mention of a Presbyterian minister.

Second, Mary and Yves might have been displeased their daughter would live in Richmond. Perhaps Therese's parents had visions of their seventeen-year-old daughter remaining single longer or, when she did marry, she might live in Norfolk where they could see her frequently. Having to travel the ninety-five miles between the Norfolk and Richmond became easier as steamships and railroad travel became commonplace; however, as parents and grandparents agree even today, any distance beyond a few miles limits visits with children and grandchildren.

Family members did make prolonged visits to Therese's home in Richmond, according to census records. Notations about these visits will appear at appropriate intervals in Therese's story. Therese will be married for the rest of her life to her beloved Tony. Almost sixty years will pass as their lives fill with children, grandchildren, their extended family, and the people of Richmond society. They were busy, involved people.

Therese moved to Richmond where Tony continued working in the family business and they started their own family. They lived with Tony's parents initially or perhaps they rented a home. Therese did follow, however, a tradition of women in her era, when she went home to her mother to have her first baby. The 1850 census taken in Norfolk includes Therese in the home of her parents. She had given birth to her first child, Marie Antoinette. The baby, not mentioned in the census, was apparently sleeping in another room and overlooked by the census taker. Therese could also have chosen to be in Norfolk because of a trip Tony made with his brother, Victor, during that timeframe.

The same month Tony and Therese were married, Tony joined what the newspaper called "a band of youngsters" on a trip west to search for gold. According to oral family history, and an account in Tony's obituary, Tony and the other young men went to California for eighteen months. I wonder what Therese thought of Tony's gold rush adventure. She left no clue. Did Tony even know she was pregnant when he left? He must have been surprised to return and meet his little daughter. Tony must have returned home by June 1851, because their second child was born in March 1852. His wanderlust apparently satisfied, Tony began working in the family business, and creating a home on the block where other family members lived as their large, influential family took its place in Richmond society.

In about 1853, Tony and his brother, George, purchased adjacent lots. They built two townhouses joined along one side where they or their surviving family members lived for half a century. Used like a name, the family spoke of each home by the address number. The home of Tony and Therese was 203. Although torn down before I visited the area, Cousin Grace of my father's generation graciously provided a description of 203 as follows: "It was a large, three-story, red brick townhouse with out-buildings in the back. The outbuildings consisted of a carriage house, a kitchen, and a laundry. The front entrance had a white column on either side of the door."

Although I don't have descriptions of the interior of Therese and Tony's home, descriptions of nearby homes provide insight. Those homes featured parquet floors, lofty ceilings, and beautiful staircases. Other family members built or owned homes on the same street and on nearby streets. Having relatives in close proximity must have been practical and comforting. Tony was one of eleven children, three of whom had died before Therese and Tony married. The men of Tony's family were actively involved in civic associations and the arts as well as maintaining their church and business connections.

Therese had a full social calendar. Fortunately, even during times when money was limited, she was able to afford help in her home. After Marie Antoinette's Norfolk birth on August 15, 1850, Therese's first son was born in Richmond on March 19, 1852, and named Charles Yves to honor Therese's youngest brother (Charles) and her father (Yves). As we'll find so often throughout Therese's life, happy events mingle with tragedy. In September 1852, when Therese had her hands full with an infant and a toddler, a two-year-old niece died. Less than a year later, they lost a six-year-old nephew.

Happy events returned on August 21, 1854, when Therese gave birth to her third child and namesake, Marie Therese "Renie." Renie never married and lived to the age of seventy-eight. The family appears to have survived 1854 with no deaths, while two hundred fellow Richmond residents died of cholera caused by impure drinking water. Those who drank from uncontaminated wells and springs survived. Since the family business involved plumbing, they understood water quality issues.

Tragedy returned on March 29, 1855, when death visited Tony and Therese in their home. Their five-year-old first child, Marie Antoinette, died. I'm so grateful to Cousin Linda who preserved a letter written by Tony's younger brother, Victor, a week after little Marie died. Victor, who was attending the Virginia Military Institute at the time, wrote about the passing of their little niece in a letter dated April 6, 1855, to his older sister, Virginia. "Oh, how distressing it must have been to Tony and Therese," he wrote, "particularly as it was so very sudden, they did not have time to prepare themselves for it."

Therese gave birth to her fourth child, Vincent Walter, on July 11, 1856. My great-grandfather's first name honored Tony's older brother, Vincent. Initially, I had no idea what prompted the use of the middle name Walter. When I learned Therese favored the works of Scottish historical novelist and poet, Sir Walter Scott, I had my answer. Called Vincent and never a novelist, Therese's son did grow up to thrill people with his spoken words.

At five in the afternoon on December 26, 1856, Tony's mother died. Catherine, who had been in excellent health, was reportedly engaged in performing light duty in her kitchen when she fell to the floor and expired soon after, despite the efforts of two physicians.

On December 27, 1858, Therese and Tony's fifth child, Harry Percy, was born. Tony and Therese marked their tenth wedding anniversary in October 1859. During their first married decade, five children were born and they lost Marie Antoinette.

On December 15, 1859, their nephew died at the age of eleven months. Little George was the son of Tony's brother George and his wife, Mary Jane. Because Tony and his siblings built homes in close proximity, Therese knew this baby well for the little darling's short eleven months of life. Therese, having experienced the loss of a child herself, must have understood Mary Jane's pain.

Tony's brother, Eugene Victor (twenty-six), who five years earlier wrote the letter about Marie Antoinette passing away, died of scarlet fever on July 20, 1860. Victor's death was perhaps unusually difficult to endure, because Victor died in France where he was attending medical school.

The 1860 census records include Tony and Therese living in their home at 203 Cary Street in Richmond. Thirty-six-year-old Tony's occupation was plumber and gas fitter. Therese was then twenty-seven, keeping house, and had four children ranging in age from eight to one. The 1860 census illustrates Therese's habit of welcoming her Norfolk family members into her home. The household Tony headed that year included Therese's sister, Mary Eliza, and her brother, Charles. Perhaps Therese was lonely for her siblings or the situation in her household was more beneficial for them than their Norfolk home. A combination of the two reasons was probably accurate, because her brother Charles was working as an apprentice tinner (undoubtedly in Tony's family business), and Mary Eliza soon met and married her future husband in Richmond. Before 1870, Mary Eliza married a man the family called her *railroad man,* and was the mother of four daughters.

The fate of Therese's brother, Charles, was quite different. 1860 was Charles's last appearance in a census. By 1870, Charles and six hundred thousand Americans alive prior to the American Civil War were gone. At the end of 1860, they must have anticipated trouble yet had no way of imagining the devastation the next five years would bring. The experience of losing her eldest brother in the war with Mexico must have given Therese reason to worry when her three remaining brothers enlisted in the Confederacy. As I thought of those young men enlisting, I recalled the many Confederate men whose uniforms were homemade. Yves, as a tailor, could have his own shop create the uniforms for his sons to wear to war. What a sad task for a father who had lost his eldest son in the Mexican War.

In keeping with naming habits of the era, when Therese and Tony welcomed their sixth child on May 27, 1861, they named him Eugene Victor (called Victor) to honor the baby's uncle who had died in Paris the previous year. Before Victor was a year old, Therese and Tony's son, Harry, died on March 17, 1862, at the age of four. Infant Eugene, son of Tony's brother, Paul, died on July 10, 1862.

On August 23, 1862, Therese's brother, Robert, died in battle at Warrenton Springs, Virginia. While the war raged on and bread shortages caused riots in Richmond, Therese gave birth to her seventh and last child, Robert, on October 10, 1862. Robert's name honored Therese's recently deceased brother. On October 29, 1862, their nineteen-month-old niece, Bessie Lee, died.

The family patriarch, Antoine, died at the age of seventy-eight on January 4, 1863, after a long and painful illness. Since his wife had passed away seven years earlier and he was so ill, the family must have been relieved when his suffering ended. Antoine had been the force behind the creation of a booming business.

Tony and his siblings took over a thriving business and inherited their father's estate, including portraits of Antoine and Catherine. Painted in 1840, Catherine appears as a lovely, dark haired woman; handsome Antoine, displaying a roguish smile, sports a gold earring. Cousin Sue rescued the deteriorating paintings and paid for professional restoration so we could enjoy those faces Therese and Tony knew so well.

By 1863, the war created severe food shortages. The price of butter in 1863, for example, had risen from $1.00 per pound to $2.50. People were hungry, afraid, and suffering from loss of men to the war. Women struggled to feed and protect their families. By the end of the war, food was extremely scarce and prices were out of sight. For a pound of each commodity, prices had risen to fifty dollars for coffee, twenty dollars for bacon, and twenty-five dollars for butter. Flour had risen to $1,200 a barrel. Fortunately, the family enclave on Cary Street included vegetable gardens and fruit trees.

Lack of knowledge about the spread of infection often meant the most superficial wound could be fatal to people treated in a hospital. Wounded soldiers often faired better at home. When Therese's brother, Charles, was wounded in 1863, Therese nursed him in her Richmond attic. Unfortunately, even her valiant efforts failed, and Charles died of wounds and disease on July 10, 1863. Perhaps hastened by the stress of war and loss of his sons, Therese's father passed away in Norfolk on October 18, 1863.

Therese knew where to look for strength to survive the difficult times. On September 18, 1864, she made a note that she had "commenced to read the Bible through" then added, "may the Holy Spirit guide and instruct me" followed by her initials.

Life in Richmond during the last year of the war was horrific. Not far from the family residence, prisoners of war incarcerated in a tobacco warehouse endured deplorable conditions. They were hungry, cold, and some were insane. As prisoners died, dogs, hogs, and rats fed on their naked bodies stacked in the street awaiting burial. The stench must have been overwhelming.

During the war, Tony served in the Home Guard in Richmond. The family business used their metalworking skills to provide hardware to the Confederate army, and provided bolts, washers, nails, and padlocks when the Confederacy outfitted the ship Yorktown for war.

General historical versions of final events in Richmond as the war ended in April 1865 appear in books and documentaries. Amazingly, I learned Therese kept a journal in April 1865 that survived in the care of her descendants. The journal entries capture significant events in the life of my family and of our country. We'll read Therese's words as she recorded them with my parenthetical comments where necessary to understand her meaning. I won't interrupt the flow of her words to explain historic events. You'll be reading one woman's views as she lived through those days. She didn't know what was happening beyond what she could see and hear.

April 1, 1865. Terrible excitement in the city. We dread an invasion.

Sunday, April 2. After standing guard all night, Tony went late to church and returned sick. Hearing all kinds of rumors. In the evening, Harry Herbert came over from Petersburg wounded, says dear Tom (her brother) *is a prisoner. We are very anxious. Retire late. Lum Brett arouses us with news of the intended evacuation.*

Monday morning, April 3, before day, a terrific explosion. We jump up and discover a fire. After an hour or so another terrific noise and Lum came to tell us that he was left. Tony had been gone for some time. After hurriedly dressing Bob and Vic (younger sons), *I went down. No language can describe the horror of that day. Tony struggled in vain. The store was burning and while he toiled down below* (down by the wharf) *the horrible element* (mob) *threatened us up here* (at their home). *The United States troops have taken possession of the City. The Negroes are frantic. Plunderers pass in advance and homeless wanderers in tears and anguish pass, not knowing where to turn from the accumulating horrors. Memory sickens at the recollection* (she's writing at the end of the day or perhaps the following day). *Harry Herbert and Lum are still with us. A week of intense excitement has passed and another dear Sabbath* (Palm Sunday) *has arrived. We have enjoyed the precious privilege of meeting in our Sanctuary once more.*

Monday, April 10, was ushered in by a salute of two hundred guns. Lee has surrendered. Sad faces meet us at every turn. When will it end?

Monday, Tuesday and Wednesday pass amid a whirl of excitement and despondency. Thursday rain. Friday, I shall leave everything and try to go to church at least once this holy week. I went to church (Catholic) *and it was closed. Alas for freedom. In the afternoon, I accompanied William Morrison to a prayer meeting at his Centenary* (Methodist) *Church.*

Sunday, April 16, Easter Day. We had hoped to have communion in our own church. It was still closed and I attended Centenary again. Two weeks since the fall of Richmond and Tony and Vint (her husband and 9-year-old son, Vincent) *and Charlie* (13-year-old son) *are just commencing to clean bricks and try to put up a frame house to begin the world over. We have tried to be as economical as possible and now we are alone once more.*

I shall try to help Tony all I can. We must not despair if we have health and patience. With trust in God, all will yet be well.

Lincoln has been murdered. Heaven knows if we must suffer for this atrocious crime of which we are innocent."

Therese was doing more than keeping a journal. Therese gave her nine-year-old son, Vincent, a bucket of water and a dipper, then sent him to offer cool water to soldiers and displaced people passing in front of their house. Even in the midst of war and absence of husbands, spiritually strong women provide Godly leadership for their children. Therese seemed hungry for the normalcy of her own church yet was undeterred when it was closed. She took her children to an available house of worship.

When Therese and her children visited Centenary Methodist Church in April 1865, they heard Rev. David Seth Doggett, D.D., a spiritual giant. When I learned Rev. Doggett had been in the pulpit, I was not at all surprised he kept his church open and continued to minister to the people of Richmond in the midst of war. Thirty-three year old Therese was listening to a man twenty-two years her senior. Rev. Doggett had the maturity to communicate calm, hopeful words to his war-weary congregation and visitors that morning. Rev. Doggett's election as bishop the following year confirms what I've read about his leadership skills and spiritual stature. How comforting for Therese to hear a sermon by a pastor said to be an eloquent and gifted speaker. Rev. Doggett was also a great scholar of the Bible who devoted himself to preaching the Gospel. His sermons were reportedly varied presentations of core truths that held the interest of his congregation without unnecessary frivolity. Therese, a woman who read her Bible regularly, must have appreciated the comfort of hearing familiar scriptures quoted by a man so interested in Bible distribution that he served as President of the Virginia Bible Society.

Therese could have stayed at home. With the enemy surrounding her city and Tony serving in the Home Guard, Therese could have skipped church. Strong women know their help comes from the Lord and recognize the need to pause, pray, and listen for guidance. Did worship, even in an unaccustomed place and manner, give Therese and her children peace and hope? Perhaps those services gave Therese the fortitude to write the phrases "we must not despair" and "trust in God" in her journal.

With children to feed and educate, Tony's family had to re-establish their company and get back to work. Ever people of integrity, Tony and his brothers worked diligently to pay their northern suppliers. Therese and the other wives were busy stretching their meager resources to keep their households fed and clothed. Housekeeping wasn't the only daytime activity in their homes. Loss of their company building meant moving their post-war headquarters to their homes.

On May 1, 1865, a Richmond newspaper article read, "Plumbing, gas fitting, and tinning will be executed in the very best style at a moderate cost and on short notice by that well-known and deservedly popular firm whose establishment was entirely consumed in the great fire of April." The article went on to direct people to the residences of Tony and his brother, George, at the corner of Second and Cary Streets. The article described the brothers as men of sterling integrity. Life had to go on.

The war was officially over; yet in the aftermath, another loved one was lost from Therese's family circle. On May 13, 1865, Therese's last surviving brother died at Point Lookout, Maryland. Tom enlisted four days after his twentieth birthday on April 18, 1861, and served the entire war, only to face capture shortly before Lee's surrender in April 1865. Therese had four brothers who, before they could marry and have families, lost their lives serving their country. When their father died in 1863, he might have hoped for male heirs to carry on his family name. After the loss of one son in the Mexican War and the remaining three sons in the American Civil War, that hope was dashed.

In December 1865, Norfolk newspapers carried an announcement of Tom's funeral. His remains were arriving from Maryland's Point Lookout Prison for a December 19, 1865, funeral at the Presbyterian church, followed by burial in the family plot. Grief was rekindled as the family laid their last young man to rest. In addition to the inscription on the family obelisk marking his grave in Norfolk, Tom's name now appears on a monument at Point Lookout at Scotland, Maryland, honoring the men who died there.

On August 26, 1866, Ida, a daughter of Tony's brother George, passed away. Ida was a baby when Therese arrived as Tony's bride in 1849. To Therese, who had watched Ida grow up next door, the loss of her niece must have been nearly as difficult to bear as losing a child of her own.

Therese turned to the source of her strength and penned a prayer on the inside cover of her little prayer book. Because it is so clearly a prayer for her husband, I've shared it with many brides-to-be. Strengthened by her prayers, Therese's marriage lasted fifty-nine years. Therese wrote the prayer near a note dated August 1867, so I associate the prayer with that era in the chronology of Therese's life. It reads:

> *O, Lord Bless and preserve that dear one whom thou has chosen to be my husband. Let his life be long and blessed comfortable and holy, and let me also become a great blessing and comfort unto him; a sharer in all his joys and a refreshment in all his sorrows; a helper to him in all accidents and changes in this world. Make me forever amiable in his eyes and forever dear to him. Unite his heart to me in the dearest love and holiness, and mine to him in all sweetness, charity, and compliance. Keep me from all ungentleness, discontent, unseasonableness of passion and humor. Make me humble and obedient, useful and observant that we may delight in each other according to Thy blessed word and may both of us rejoice in thee having our portion in the love and service of God forever and may we be eternally united through Jesus Christ our Lord. Amen, Therese*

Therese was quite young when Tony proposed marriage so perhaps he caught a glimpse of the woman she would become. After reading her prayer, I believe he chose well. Therese clearly knew her role was too difficult to fly solo. She needed and wanted God on her side.

Therese frequently hosted lengthy visits of relatives and friends. In the autumn of 1870, Therese welcomed her brother-in-law, C. P., and seven-year-old niece, Maude. Her sister, Mary Eliza, had remained behind in their North Carolina home with children who were too young to travel.

In October 1870, Richmond experienced severe flooding. The family business was located in a flood-prone area where boats were required to remove goods from flooded warehouses. Fortunately, the family homes were located at higher elevations. Water issues didn't end with flooding. Post-war Richmond experienced a drastic population increase and the need for water eventually necessitated the construction of pump houses to carry water to reservoirs from the James River. The only water available for baths, washing clothes, cooking, and drinking was murky with silt from the river. Disease was a recurring issue arising from unclean water. Impure water and tainted milk (not inspected at the time) spurred rampant typhoid fever in 1870. Therese somehow managed to keep her children alive. On a happier note, when time and money permitted, the family enjoyed musical performances and plays presented in Richmond in the early 1870s.

On June 1, 1871, Therese's mother, Mary, died in Norfolk. Mary's death left Virginia, Therese's younger sister, in charge of the family estate. Virginia's well-documented mental instability undoubtedly became a concern for Therese, and made life miserable for the youngest sibling, Fanny. Fanny moved to Richmond where she met her future husband. Virginia, perhaps jealous, tried to interfere with Fanny's marriage plans. We all know the stress a troubled family member adds to our lives.

Tony, his brothers, additional family members, and employees rebuilt the family business base then borrowed funds to construct a new company building. The new building, completed in 1874, allowed space to become one of the largest plumbing, gas fitting, and metal working businesses in the South. As any wife of a husband striving to build, rebuild, or expand a business knows, Therese had a role as Tony's supporter and encourager.

On September 9, 1875, Therese and Tony celebrated the marriage of their eldest child, Charles, to long-courted sweetheart, Emma. They knew Emma well even though she had lived for a time in Staunton, Virginia. Therese had invited Emma to visit their Richmond home in 1871, and when Charles mentioned the invitation, it sounded as though Emma had been there on previous visits. Thanks to letters written by Charles to Emma during their courtship and graciously provided by their great-granddaughter, Linda, I feel as if I know this couple well. Charles's detailed accounts of life in his family's home provide many clues about Therese's daily domestic life.

Written between 1870 and 1875, Charles's letters explain he was working for his father and uncle in the family business. When he was eighteen years old, he began work at six o'clock in the morning and worked until seven o'clock in the evening. He walked the streets of Richmond visiting customers in rain, cold, and heat. He then went to Old Dominion Business College (later Richmond College) to study Commercial Mathematics and

Bookkeeping. After his classes, he walked a mile and a quarter to their home arriving at about ten o'clock. He then ate oyster crackers and milk while he studied in his room and wrote letters to Emma. No wonder he was frequently ill.

Charles described the difficulty of bathing. Even in a home where plumbing fixtures and installation expertise put them in the best position of anyone in the neighborhood, having enough hot water was a major issue. A tank by the kitchen stove generated hot water for bathing and fed hot water into pipes and coils used to heat upstairs rooms. That system held a limited amount of water. One evening, Charles wrote that his room was cold because all the members of the household had bathed.

Lighting was an issue as well. Flooding often caused an interruption to gas service. Therese's home, outfitted for gas lighting, required the use of substandard supplemental oil lamps and candles when gaslights failed, according to Charles. Gaslights, restored after being off, produced such brilliant flickering light (caused by air in the gasometers, Charles explained) he feared eye damage. Therese, an avid reader, also had to deal with the lighting issues Charles described, or put off her reading until daylight hours.

Charles tried to shield Therese from worry. When he was experiencing chest pains, he refused to tell his mother. In a letter to Emma, Charles wrote that his father was ill. Because his mother was uneasy about Tony, Charles hid his illness. At work, Charles kept a co-worker nearby who could send for a doctor, if necessary, and told Emma he was willing himself well.

Perhaps the greatest insight about the family's financial difficulties after the war appears in the text of one of Charles's letters when he explained delaying his marriage because of financial concerns. His father, he explained, had lost $150,000 in 1865. Prior to the war and the burning of Richmond, Tony had been a wealthy man. Charles felt his parents gave him anything he requested before the war. After the war, however, Tony did not have the funds to help Charles start a business of his own.

During the difficult reconstruction era, prices on supplies from the northern states apparently continued to climb. As Tony and his family attempted to rebuild their business, price fluctuations made returning to profitability extremely difficult. They bid jobs based on current prices; yet by the time they procured materials, prices had risen. Apparently, because of their own ethics, they refused to pass the increase in price to the end customer, and they found themselves losing money on many of their jobs. Therese must have been on her knees in fervent prayer.

Charles mentions Therese numerous times in his letters. Often, he quotes his mother's opinions or comments. My favorite excerpt is from a letter dated February 7, 1871. It reads, "Last week we had the most magnificent natural curiosity here in Richmond. A crystal fountain was formed by water from one of the fountains freezing as it went up. Ma went down to see it, and she thought it was the most beautiful thing she had ever seen. You would have thought she was going crazy over it if you could have heard her talk of it." At thirty-eight, Therese unreservedly expressed joy and wonder. Taking time to admire frozen fountains probably delivers the same stress-relieving chemicals as does taking time to smell the roses.

In a January 1875 letter, Charles mentions that his mother sent his fiancé, Emma, a publication for women called *Household*. Therese, Charles explained, earned the money (presumably by selling subscriptions) to pay for Emma's subscription. Therese apparently found small ways to support Tony's financial recovery effort.

Even in times of financial difficulty, Tony and Therese educated their children. Charles wrote to Emma about his sister attending private school in the early 1870s. Tony also sent my great-grandfather, Vincent, to a private academy then to Richmond College in 1872 and 1873. In addition to Vincent's college courses, Tony arranged practical business training in Kentucky where Vincent received training beyond his father's expectations. The Methodist family with whom Vincent was staying took him to a Methodist camp meeting where Vincent reported he was "saved and received a call to preach."

When Vincent wrote to inform his parents of his decision to enter the ministry, his father was upset and insisted he return home right away. Based on Therese's spiritual views, she couldn't argue with such a calling and was probably pleased. Vincent returned to Richmond, reached agreement with Tony, and then Tony funded Vincent's seminary education at Randolph-Macon.

Meanwhile, Therese was the wife of a busy man. Tony, in addition to his business career, became an alderman for the City of Richmond. In 1876,

Tony began serving on the board of directors of Hollywood Cemetery, Richmond's historic burying ground. Tony's civic appointments signify the trust of his peers and their acknowledgment of Tony's contributions to Richmond's successes. Tony also served as a member of the board, and on the finance committee, of the Mechanic's Institute (a school for men); and served on a bank board of directors.

Tony's demonstrated leadership in Richmond's cultural life by serving as president of an association for the arts. The Mozart Association, formed in 1876 to promote classical music, initially met in Therese and Tony's parlor, and in the homes of other members. In the 1880s, the association grew to two thousand members. Tony and other prominent businessmen led and supported the association's endeavors to present musicals, Thursday evening concerts, operas, and operettas.

In 1886, the organization dedicated The Mozart Academy of Music on Richmond's Eighth Street between Grace and Franklin. Tony served on the building committee of the new facility, intended to house the Mozart Association's growing membership and activities. On opening night, January 11, 1886, Tony and Therese must have dressed in their finery to attend an elegant opera. The new venue allowed Richmond citizens to enjoy New York-based touring company performances. Tony and Therese, great lovers of music, attended musical performances regularly and enjoyed a beautiful rosewood piano in their home.

Tony's financial contributions to the Academy of Music construction created a rift that ultimately ended their immediate family's Roman Catholic affiliation. According to family lore, their Catholic priest expected Tony to dedicate all his available funds to their church, and chastised Tony for donating money to the Academy of Music. Tony and Therese became Episcopalian. Tony became a leader at Grace Episcopal Church where he served as a Vestryman and Senior Warden. Therese, who had Presbyterian, Catholic, and Methodist worship in her history, left no commentary about the change; but her journal entries point to willingness to worship wherever the gospel was preached.

Tony's business and civic duties left little time for Therese. I assume Tony, like other businessmen in Richmond, went home for the main meal of the day, called dinner, at about 3:00 p.m. After dinner, men typically returned to work for several hours, then came home in the evening and ate a light supper. I have wondered if Tony and Therese had private, enjoyable conversations over their mid-afternoon meal if the children were in school. If they weren't alone during meals, private time for the couple might have been limited to bedtime since one or both of them attended meetings or events most evenings.

Tony and Therese's first grandchild was born on June 9, 1876, when their son Charles and his wife Emma welcomed Emma Therese. Charles lived nearby so Therese saw her namesake often. Ida Estel, another granddaughter, joined Charles and Emma's household on November 6, 1877.

On February 10, 1877, Tony's brother, Vincent, died at the age of fifty-nine. On July 4, 1878, four-year-old Mary, the daughter of Tony's brother, Paul, died. When a beautiful photograph of a mother and infant came to me for identification, I realized the baby was Mary and the mother was Paul's wife, Laura. The sweet, peaceful mother-and-child happiness of the photograph reveals not a hint of the sadness that was to come.

Also in 1878, Fanny, Therese's youngest sister, married Joe, the son of Anthony's older sister, Jane. Since Fanny lived in Richmond with Tony and Therese after her parents died, it was easy for romance to blossom between Joe and Fanny, who were frequently together at family gatherings. Joe and Fanny were living in San Francisco, California by 1880 and their only child, Joseph Arthur, was born there on July 25, 1881. In 1900, they were living in Oregon, and then Fanny and Joe died before their son registered for World War I in 1918. Fanny's move west must have been difficult for Therese. Perhaps her baby sister never returned to Virginia. Fanny had delayed her marriage because of difficulties with her sister, Virginia, who apparently wasn't happy and didn't want others to be happy. Therese must have been pleased when Joe's widower father and two of Joe's sisters accompanied Joe and Fanny to California. If Therese couldn't be with Fanny, at least she knew Fanny had a family support system.

It was Tony's turn to welcome a namesake. On February 18, 1879, Charles and Emma welcomed a third child and named him Charles Anthony. Charlie lived near Tony and Therese for the rest of his grandparents' lives.

On June 18, 1879, Virginia, Therese's mentally troubled sister, died at the age of forty-two of suicide by charcoal poisoning. Left in charge of the Norfolk family estate when her mother died, Virginia behaved irrationally. Therese and Fanny considered confining Virginia in the asylum at Williamsburg when the family couldn't deal with her behavior. Perhaps Therese felt some guilt over leaving Virginia on her own. At least unhappy Virginia was finally at rest.

Tony's eldest brother and business partner, George, passed away May 8, 1880, at the age of sixty-three. A Richmond newspaper referred to George as an "old merchant and member of the City Council." The family lost many members in the next few years. Therese attended funerals for a niece who died in 1881 at the age of thirty-nine, a great niece who died in 1883 of scarlet fever at age nine, and Tony's brother, Paul, who died in 1883 at the age of fifty-seven. During an era of strict rules about proper attire during mourning, I wonder if the women of the family were perpetually dressed in mourning black.

On November 16, 1883, tragedy visited Therese and Tony when their son, Robert, twenty-one, died of tuberculosis. He was the much-loved namesake of Therese's brother, Robert. According to the 1880 census, Robert had been working as a clerk in a feed store so, for unknown reasons, he had not joined his father in the family business. By 1883, newspaper reports of death were more personalized and informative. Writing style and content of Robert's obituary indicate Therese was the author. After a simple statement giving his name, the date of his death, his parent's names, and mentioning he was the youngest son of his family, the following was printed:

> *"It is all right, Mamma; it is worse for you than for me," were his last words; and those who knew him best, know how sincerely he meant it. From his infancy up, he was a marvel of virtue, purity, and integrity, and for the last few years of his life was an earnest, consistent member of the Episcopal church. Though liberally endowed with rare talents of a high order, he was modest and humble to a fault. During his brief business career, he rose rapidly in the confidence and good opinion of his employers and business acquaintances; and the nearest to a complaint that was ever heard of him was, "He is too honest to succeed in business." His last illness extended over a period of ten months, and though subjected to much suffering, and to several severe and bitter disappointments, he bore all with a patient Christian fortitude rarely*

> *beheld; and at last, in the presence of his entire grief-stricken family, he passed quietly from this world into that higher and brighter sphere prepared by the hand of an almighty and merciful Father for such as he. May we all meet him there.*

A family milestone was celebrated 1884 when Tony and Therese's son, Vincent, began his fifty-year career as a Methodist minister. A later biographical sketch of Vincent's life stated that by the time he was an active minister, his father said he had been wrong about Vincent making a mistake to choose a career outside their family business. Tony reportedly told Vincent he had obviously made the correct choice, when he saw how his son prospered in his calling. Tony was pleased when he saw Vincent encouraging souls to turn to God, and starting churches where there had been none. Therese must have praised God.

Therese and Tony welcomed another grandchild when Charles and Emma's fourth child, Linda Josephine, was born on July 7, 1885. Linda, by all accounts, was a much-loved addition to the family. Even today, people who knew her say family members wanted to visit Linda and her sister, Ida, as often as possible. Linda is the grandmother of Cousin Linda who provided extensive information about our family.

Charles and Emma completed their family in April 1889 at the birth of their son, Robert. Therese must have been pleased when Charles created a namesake for his brother who died in 1883. Therese and Tony had five grandchildren living nearby. Perhaps they were thinking none of their other surviving children (Vincent, Victor, and Renie) were going to marry or have children. Victor and Renie did not marry.

Vincent's ministry had taken him to the Eastern Shore of Maryland where, in January 1890, he married Daisy. You read about that couple and their children in Daisy's story. Of importance to Therese and Tony, Vincent and Daisy gave them five more grandchildren between 1891 and 1906.

Did Therese ever hear Vincent preach? Surely, she did. Vincent did pastor a Richmond church during the year his mother passed away. I hope he arrived in time for Therese (if she were well enough) to attend his services in Richmond before she died. Perhaps he preached revivals in Richmond churches, and Therese had opportunities to attend those services. She may have read glowing reports of his ministry published as early as 1901. Even if they didn't hear him preach, his parents knew he was following the path God had chosen for him.

Tony's involvement at Hollywood Cemetery grew. On January 2, 1889, after serving as a board member since 1876, Tony became the fourth president of the Hollywood Cemetery management company. Tony visited the cemetery daily, and became involved in every minute detail. Tony's work must have pleased the board. He served without compensation until April 1894; when the board voted to pay him $600 a year, and gave him a horse and buggy. I include Tony's many civic activities as part of Therese's life for a reason many women understand. All the time Tony spent working on the cemetery took time away from their personal lives.

Events related to Therese's troubled son, Victor, came to a head in 1890, culminating in Victor leaving home. I first heard Victor left Richmond because he had tuberculosis. If Victor was ill, perhaps he had lived too close to his brother, Robert, who died of tuberculosis. He could have sought healing by going west to live in fresh, outdoor air. Cousin Grace provided another explanation that better matches other facts. Grace understood Victor had a drinking problem. According to Grace, Therese and Tony told Victor he needed to address his drinking problem or leave. Newspaper clippings revealing Victor's behavioral tendencies support the drinking problem scenario. Ultimately, he chose to live on his own terms.

A newspaper report indicated that on August 3, 1886, Victor and two other men finished (building or rebuilding) a handsome yacht, then left Richmond for an extended sailing trip. While his siblings were working hard, Victor was playing. The timing of Victor's yacht trip seemed even more inappropriate when I learned that shortly before he was to depart, the family business won a large contract for installing the steam system at an asylum in Marion, Virginia. The size of the contract necessitated subcontracting to other Richmond firms. Rather than sail off into the sunset, Victor could have offered to support his family's business. After his father funded two years of college, Victor apparently hadn't chosen a vocation.

On August 11, 1887, a public carriage driver reportedly stole thirty dollars from Victor. The driver, initially arrested, jailed, and refused bail, was acquitted when the case came to trial on March 20, 1888. Did Victor lose his money while under the influence of alcohol? Did he incorrectly blame the carriage driver? Perhaps Victor was unable or unwilling to live as his parents expected. Whatever the reason, Vic (as he was called by family and friends), went west. Leaving Richmond in about 1890, Vic went to Colorado then to Idaho. Thanks to Cousin Paul and his wife, Joan, who commissioned an Idaho researcher, we know a great deal about Vic's Idaho life.

Vic, the mountain man, as we sometimes refer to him, worked as a miner, trapper, and photographer. He acquired business associates and cultivated friends, but never married. He worked his trap lines alone in the Idaho wilderness, and lived in little cabins he built for himself along the route. He left his mark. An Idaho creek bears his family name and two mountain names honor Vic's horses. He remained in Idaho until his death a decade after Therese passed away. There is no evidence Vic corresponded with his parents after leaving home. Toward the end of his life, Vic corresponded via letter with his brother, Vincent, who was responsible for Vic's estate when he died. Vic never returned to Richmond.

Even if Vic was a challenging child to parent, Therese must have been sad when her son left home. Tony was probably disappointed Vic did not join the family business. Poor Charles, who had been working so hard since his teen years, was the only member of Tony's immediate family to spend his career with the family firm. Charles's last hope for help from his siblings disappeared when Victor headed west.

By 1892, Tony's company was the largest business of its kind south of Baltimore. Prosperity returned, allowing Therese to live comfortably again financially.

Therese, who mentioned reading her Bible in the late 1860s, was no stranger to the *Word*. According to notes in her Bible, she again "commenced to read the Bible through" on June 26, 1895, and finished on August 29, 1896. She began reading her Bible again on August 29, 1896, and finished on September 15, 1897. Therese knew where to turn for wisdom and strength. Her example prompted me to return to the practice of reading through the Bible annually.

As the century ended, Tony and Therese reached their golden wedding anniversary in October 1899. If they had no formal celebration, they must have spent a moment reflecting on their fifty years of marriage.

The makeup of Therese's household in later life illustrates her lifestyle of hospitality. In 1900, the census record for their dwelling includes Therese's niece, a young woman named Cora. Cora's mother had been Mary Eliza, Therese's younger sister. By the time Cora was living with Therese, her parents had apparently passed away.

Tony was embroiled in a lawsuit in August 1902. During a hotel construction project, another company purchased supplies from Tony's firm, and then failed to pay their $10,000 bill. In April 1903, the case remained in court and review of evidence continued without resolution.

In October 1902, Tony tried to stop the shooting of rifles and pistols near Hollywood Cemetery. The practice, he claimed in a formal complaint, posed a danger to visitors at Hollywood.

The evening of July 23, 1903 was stressful for Therese. A streetcar hit Tony's horse and buggy. Then seventy-nine but quite active, he was en route to Hollywood Cemetery. Tony complained that the streetcar failed to ring a bell. The streetcar smashed the buggy, and badly injured the horse. Although thrown from the buggy and severely bruised, Tony escaped serious injury.

A fifty-three-year-old nephew, who had lived next door all his life, died in August 1903. Perhaps the greatest loss for Therese occurred on March 28, 1904, when her sister-in-law, Mary Jane, passed away. Therese and Mary Jane married brothers, raised their families a few feet apart, and shared life's happy as well as sad times for half a century.

Therese must have enjoyed seeing her grandchildren happily married. In 1904, Emma Therese married. Emma's sister, Linda, married in 1905.

Tony continued to work at his business and in support of community activities. In September 1905 (while still serving as president of the Hollywood Cemetery Company) the American Association of Cemetery Superintendents invited Tony to Washington, D.C. where he received an honorary membership for his many years of effort on behalf of his city's cemetery. It must have pleased Therese for Tony to receive peer recognition for his many years of service.

Therese's nineteen-year-old granddaughter, Ida, almost drowned at Virginia Beach. Ida was swimming in the ocean on July 27, 1906, when surf swept her under. Another young woman pulled Ida toward the shore where two men pulled her from the surf. Ida, unconscious, was resuscitated by rolling her on a barrel (to force out water) while several other swimmers were rescued. Although Ida was very ill for a day after the rescue, she recovered and lived until 1943. The story must have terrified Therese who worried about her loved ones.

Mothers are never prepared to outlive their children. Tony and Therese lost their eldest son, handsome fifty-four year old, Charles, in January 1907. Tony must felt the loss of Charles keenly since Charles and Tony had labored side-by-side for decades during the lean and fat years of the family business. Charles, like his father, was a civic and church leader as well.

At ten o'clock in the evening of on Sunday, July 19, 1908, a policeman noticed a fire in Tony's company building in downtown Richmond. While firefighters, one of whom was mortally injured, managed to save valuable surrounding buildings and businesses, Tony's building and contents were damaged in excess of $60,000. At least this time, unlike losing their business to the 1865 fire, insurance would help the family recover their investment. The stress of losing the family business again and her concern for Tony's ability to deal with the loss may have hastened Therese's death.

On August 5, 1908, Therese died at the age of seventy-six. Her burial took place in the family plot at Hollywood Cemetery. Tony, eighty-five years old and without Therese for the first time in fifty-nine years, must have been lonely. Not quite nine months later, on May 1, 1909, he was preparing Hollywood Cemetery reports for the annual meeting when he suffered a heart attack, collapsed at his desk, and died at 9:40 p.m. Two days later, his burial took place beside the grave of his beloved Therese.

Who was Therese and what legacy did she pass along to those who came after her? Their descendants say Therese and Tony enjoyed mutual love and admiration. Did Therese pass along her faith in God? We don't know whether Victor worshiped in the Idaho wilderness. Her son, Robert, was a man of faith, according to the obituary written at the end of his brief life. We know the other children attended church and we have documented evidence of the dedication of two sons. Charles was active in church leadership as a layperson while fulfilling his father's wishes by joining the family business; and Vincent chose Christian ministry as his life's work.

Therese left evidence of her conversations with God when she penned the prayer I now call *the wedding prayer.* We know she was a disciplined student of the Bible, and read it through at least three times. Sorrow and difficulty were no threat to a woman who placed her trust in God. Faith must have seemed all she had left when beloved siblings and children were lost to death, war raged around her home, and money was gone. Yet during her lifetime, when smiling in photographs wasn't the fashion, Therese wears a wistful smile in every photograph. Through life's trials and sorrows, evidence indicates Therese drew strength from her faith.

We know Therese to be a woman of strength and character. A daughter of parents who blended genetic material from France and the British Isles, she was a physically attractive woman with light blue eyes and brown hair. Her face was oval, and softened only a bit over the years. My grandmother, who was born when Therese was sixty-one, described Therese as a small, plump woman with tiny hands and feet.

Long before I knew anything about Marie Therese, I felt connected to her through her first name. Marie became the first name of her granddaughter, later my grandmother, Marie. I inherited Marie as my middle name from my grandmother. Having learned how Therese lived her life, it is an honor to be her namesake.

The meaning of Therese's name is a prophecy for her life. Therese means *she who reaps.* What a fitting description for a woman who knew the importance of sewing the seeds of God's word in her life by reading her Bible. She knew sowing produced the best yield when followed by weeding (ongoing Bible study and confession), and watering (prayer and worship). Now she's reaping a bumper crop of eternal reward for her spiritual investment. Let's follow her example and leave instructive notes in the margin of our prayer journal or Bible for future generations.

Therese welcomed her granddaughter, Marie, into her Richmond home the winter of 1907 because she loved having young people around. Half a century later, Marie welcomed me to her Richmond home, and spent time telling me about her beloved grandmother, Therese. Perhaps Therese taught Marie the true meaning of the word *devotedly* that Marie wrote in closing her letters to me. Therese was indeed a woman of great devotion. She cultivated her relationship with God, focused much energy toward loving and helping her husband, and actively nurtured relationships within her immediate and extended family as well as among members of her society.

Thanks to Therese, and the spiritually strong women in your own family tree, we are encouraged to be women who recognize and seek the greatest source of strength: God. Because she asked (in prayer), God gave Therese (a woman with tiny hands and feet) a big, strong, muscular spirit to carry the heavy burdens of life. If we could see a picture of her spiritual condition, we'd see she's wearing armor. Every time she read her Bible through, she was putting on the armor of God, as Ephesians 6 teaches us. What happens when we wear Godly armor? The Ephesians passage goes on to assure us that when life's battles end, we will be *standing up*. Not defeated. Not overwhelmed. Victorious.

When I think of dear great, great-grandmother Therese, the last words of the 1883 obituary of her son, Robert, come to mind, "May we all meet him there." I imagine Therese penning those words herself and, without knowing us, she included all future generations of her physical family and all believers in our universal spiritual family in her word *we*.

Christians often speak of standing before God in eternity. I believe Therese initially stood before God then assumed the posture she knew well in this life: she's on her knees. She's facing God's throne and smiling. Therese is seeing the Lord she loved here and in whose presence she rejoices there: Jesus.

Encouraging Words

My stories end, your quest begins.

Did you find reading the stories of my ancestral women exhausting? Imagine living their challenging lives without modern conveniences. Thankfully, they succeeded. I am living proof.

I hope you soon discover the names of the amazing women in your own family tree, research their lives, and learn of their faith. Allow their lives to encourage and instruct you. May their decisions positively inform your choices.

If you're not comfortable beginning research on your own, gather your female kin or your girlfriends and form a *Beyond DNA Club*. Take a genealogy class together. Meet as a research group to encourage and help one another. Invite your sisters, daughters, and granddaughters on vacations to places where your ancestral women lived. Put flowers on their graves. Thank them.

When we've become acquainted with our spiritual ancestors, we'll want to spend eternity in God's presence with them. How do we guarantee our eternal destination? What does God tell us in His Word?

Open your Bible to the Gospel of John. Read the first three chapters. All will become clear. The path to salvation we take today is the same path our foremothers walked before us, and our descending daughters will walk in the future. God never changes.

Build your new spiritual muscles or tone your old ones. Seek God by reading your Bible. Find a church home where you will hear great Bible-based teaching, and where you can meet friends who will pray with and for you.

In your mind's eye, envision the scene in heaven when we begin our personal relationship with God through Jesus Christ. Can't you imagine our female ancestors clapping, cheering, shouting, dancing, and singing with joy?

May your children's children call you a strong spiritual woman who made the world a better place. May your shining example light their path toward heaven.

Blessings!
Selena

About the Author

Two years before men walked on the moon, Selena Post began her career in aerospace at Houston's Johnson Space Center. Three decades later, she had specialized in computer security and managed a program that received national recognition. With each job assignment, her tasks increasingly included motivating coworkers toward excellence. Career achievements, even in rocket science, benefit mankind in this earthly, temporal life where everything will end.

A Christian since the age of ten, she sought a unique venue to inspire women toward more eternal goals. Women can influence countless generations by attaining personal spiritual strength, then providing positive spiritual leadership for their children and grandchildren.

Seeking role models for her own spiritual journey, she discovered evidence of faith among her ancestral women. By studying the lives of her ancestors, and seeking evidence of their worship, Bible study, and prayer habits, she learned how they endured life's challenges. By evaluating the evidence of faith in previous generations, Selena found successful women to emulate. Inspired by what she found, Selena insightfully motivates women to supplement traditional genealogical research to identify their own spiritual history.

With this book, documenting stories of fifteen women in her own ancestry, she describes an engaging process for women to learn of the spiritual giants in their past, then provide spiritual guidance for the eternal benefit of future generations. Selena provides insights to facilitate genealogical research while encouraging spiritual stamina among her readers.

Selena has retired from aerospace four times, and lives in Houston, Texas, where she attends Second Baptist Church.

CPSIA information can be obtained at www.ICGtesting.com
Printed in the USA
LVOW090538181211

259932LV00002B/28/P

9 781462 705030